Purposeful Design

SCIENCE
THIRD EDITION

GRADE 1

purposeful design
publications
A Division of ACSI

Colorado Springs, Colorado

© 2003, 2014, 2020 by Purposeful Design Publications
All rights reserved. First edition 2003.
Second edition 2014
Third edition 2020

Printed in the United States of America
28 27 26 25 24 23 22   2 3 4 5 6 7 8

Elementary Science, Grade 1 – Student Edition
Purposeful Design *Elementary Science* series
ISBN 978-1-58331-732-7, Catalog #203011

No portion of this book may be reproduced, stored in a retrieval system, or transmitted, in any form or by any means—mechanical, photocopying, recording, or otherwise—without prior written permission of Purposeful Design Publications.

Purposeful Design Publications is committed to the ministry of Christian school education, to enable Christian educators and schools worldwide to effectively prepare students for life. As the publisher of textbooks, trade books, and other educational resources, Purposeful Design Publications strives to produce biblically sound materials that reflect Christian scholarship and stewardship, addressing the identified needs of Christian schools around the world.

References to books, online resources, and other ancillary materials in this series are not endorsements by Purposeful Design Publications. These materials were selected to provide teachers with additional resources appropriate to the concepts being taught and to promote student understanding and enjoyment.

Unless otherwise noted, all Scripture quotations are taken from the THE HOLY BIBLE, NEW INTERNATIONAL VERSION®, NIV ® Copyright © 1973, 1978, 1984, and 2011 by Biblica, Inc.® Used by permission. All rights reserved worldwide.

The image of MyPlate on page 237 courtesy of the USDA Center for Nutrition Policy and Promotion.

The Lego Group does not sponsor, authorize, or endorse this textbook.

Purposeful Design Publications
731 Chapel Hills Drive • Colorado Springs, CO 80920
800-367-0798
www.purposefuldesign.com

Table of Contents

Foundations

Chapter 1: Introduction to Science
1.1 Science and Scientists 5
1.2 Scientific Investigations 7
1.3 Science Tools 9
1.4 Recording Data 11
1.5 Engineering and Technology 13
1.6 Investigate: Engineering Design Process 15
1.7 Chapter 1 Review 17

Life Science

Chapter 2: Animals
2.1 Living and Nonliving 23
2.2 Animals Have Needs 25
2.3 Animals Have External Parts 27
2.4 Animals Can Be Grouped 29
2.5 Animal Babies and Their Parents 31
2.6 Investigate: Body Coverings 33
2.7 Chapter 2 Review 35

Chapter 3: Mammals
3.1 What Mammals Are 39
3.2 Mammals Have External Parts 41
3.3 Life Cycle of a Beaver 43
3.4 Mammal Babies and Their Parents 45
3.5 Mammals of the Woodlands 47
3.6 Investigate: Mammals Change Their Environments 49
3.7 Chapter 3 Review 51

Chapter 4: Fish and Birds
4.1 What Fish and Birds Are 55
4.2 Fish Have External Parts 57
4.3 Birds Have External Parts 59
4.4 How Fish and Birds Detect 61
4.5 Life Cycles of Fish and Birds 63
4.6 Investigate: Camouflage 65
4.7 Chapter 4 Review 67

Chapter 5: Plants
5.1 Plants Have Parts 71
5.2 Life Cycle of a Plant 73
5.3 What Plants Need 75
5.4 Plants Are Useful 77
5.5 Plants Protect Themselves 79
5.6 Investigate: Seed Growth 81
5.7 Chapter 5 Review 83
Unit Connect 85
Design 87

Physical Science

Chapter 6: Movement and Sound
6.1 What Makes Things Move 99
6.2 Ways Things Move 101
6.3 Investigate: Changing Motion 103
6.4 Making Sound 105
6.5 Nature of Sound 107
6.6 Investigate: Testing Sound 109
6.7 Chapter 6 Review 111

Chapter 7: Light
7.1 Seeing Light 115
7.2 How Light Travels 117
7.3 What Light Passes Through 119
7.4 Reflection 121
7.5 Bending Light 123
7.6 Investigate: Communicating with Light 125
7.7 Chapter 7 Review 127
Unit Connect 129
Design 131

Earth Science

Chapter 8: Objects in the Sky
8.1 Sun 143
8.2 Day and Night 145
8.3 Moon and Stars 147
8.4 Patterns in the Night Sky 149
8.5 Differences in the Day and Night Skies 151

8.6 Investigate: Shadows..153
8.7 Chapter 8 Review 155

Chapter 9: Weather and Seasons
9.1 Weather..159
9.2 Measuring Weather 161
9.3 Seasons..163
9.4 Seasons and Plants 165
9.5 Seasons and Animals 167
9.6 Investigate: How the Sun Warms Surfaces......................169
9.7 Chapter 9 Review 171
Unit Connect...173
Design 175

Human Body

Chapter 10: Five Senses
10.1 Sight and Hearing 183
10.2 Touch 185
10.3 Smell and Taste 187
10.4 Human and Animal Senses...189
10.5 Senses and Safety..191
10.6 Investigate: Food and Senses 193
10.7 Chapter 10 Review 195

Chapter 11: Bones, Muscles, and Teeth
11.1 Bones 199
11.2 Muscles and Joints 201
11.3 Healthy Bones and Muscles...203
11.4 Teeth..205
11.5 Healthy Teeth 207
11.6 Investigate: Tooth Care 209
11.7 Chapter 11 Review 211

Chapter 12: Heart and Lungs
12.1 Heart 215
12.2 Heart and Blood 217
12.3 Lungs..219
12.4 Lungs and Breath 221
12.5 Healthy Heart and Lungs 223
12.6 Investigate: Number of Breaths 225
12.7 Chapter 12 Review 227

Chapter 13: Stomach and Food

13.1 Stomach and Eating 231
13.2 Food as Fuel 233
13.3 Where Food Comes From 235
13.4 Good Food Choices 237
13.5 Healthy Habits 239
13.6 Investigate: Greasy Potato Chips 241
13.7 Chapter 13 Review 243
Unit Connect 245
Design 247

Glossary 255
Science Journal 257

Unit 1
Foundations

By wisdom the Lord laid the earth's foundations, by understanding He set the heavens in place.
Proverbs 3:19

Marbles get their colors from sticks of colored glass.

If you could make marbles, what would they look like? Draw them in the space below.

Name _____

Science and Scientists

What does a scientist do?

Ann sees a bird with a twig.
The bird takes the twig up a tree.
What might Ann wonder?

Ann wants to know why the bird has a twig.
She asks Dad what the bird will do.
Ann watches the bird.

1. Circle what the bird will make.

Scientists ask questions and look for answers.

2. Is Ann acting like a scientist? Write Yes or No. _____

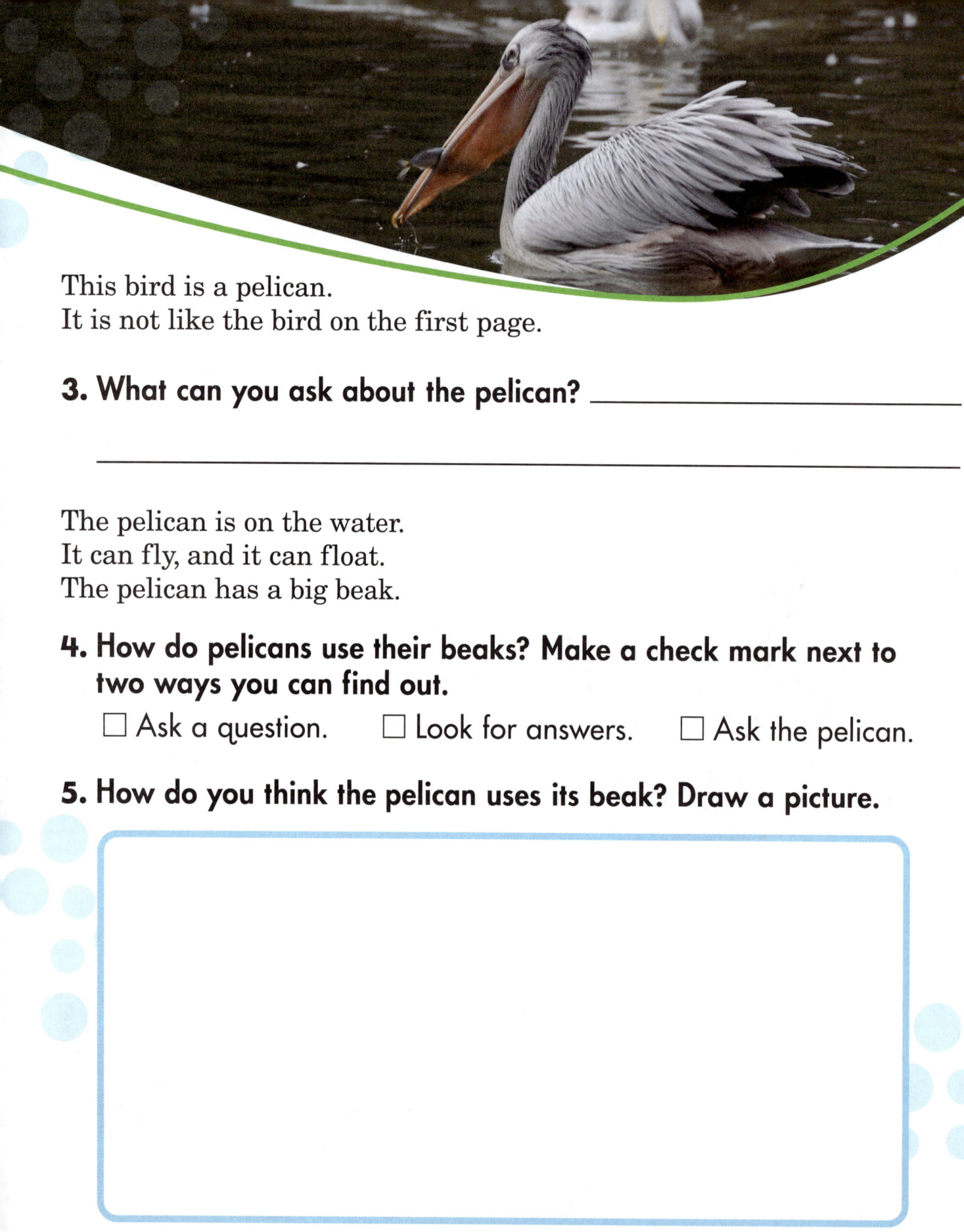

This bird is a pelican.
It is not like the bird on the first page.

3. What can you ask about the pelican? _____

_____

The pelican is on the water.
It can fly, and it can float.
The pelican has a big beak.

4. How do pelicans use their beaks? Make a check mark next to two ways you can find out.

☐ Ask a question.   ☐ Look for answers.   ☐ Ask the pelican.

5. How do you think the pelican uses its beak? Draw a picture.

You act like a scientist when you ask questions and look for answers.
God made you to wonder about and explore His world.

Name _____

Investigation 1.2

Scientific Investigations

What is the purpose of an investigation?

Step 1 Ask a Question

What do you want to know?
Will both kinds of corn kernels pop when heated?

Step 2 Make a Prediction

What do you think the answer will be?

Circle your prediction.
Both kinds of kernels **will / will not** pop when heated.

Materials
- 2 bowls
- popcorn kernels
- feed corn kernels
- measuring cup
- 2 small paper bags
- microwave oven
- oven mitt

Step 3 Plan and Do a Fair Test

How will you look for an answer?
I will heat both kinds of corn kernels.

Make a check mark for the sentence that shows the fair test.
☐ I will use the same amount of each kind of kernel.
☐ I will use a different amount of each kind of kernel.

Step 4 Record and Analyze Results

What did you discover?

Make a check mark on the chart.

	does pop	does not pop
popcorn		
feed corn		

Step 5 Make a Claim

How do the results compare to your prediction?

Make a check mark by the correct answer.
☐ Popcorn pops when heated.
☐ Feed corn pops when heated.

Step 6 Give Evidence

How do you know your answer makes sense?

I planned a fair test.
I looked at both kinds of corn kernels after heating them.

Step 7 Share Results

What can others learn from your discovery?

Make a check mark for the correct answer.
☐ Others can learn about which corn bakes.
☐ Others can learn about which corn pops.

The purpose of an investigation is to look for answers that make sense. You can investigate the world God created.

Name _____

Science Tools

Which tool should I use?

Scientists use tools.
A **tool** is an object used to do work.
Tools help find answers that make sense.
It is important to choose the correct tool.
A hand lens makes objects look bigger.

Some tools measure things.
Measure means to find the size, length, or amount of something.

 A ruler measures how long or how tall something is.

 A measuring cup measures liquids.

 A measuring tape can measure around something.

1. Circle the pictures of tools.

 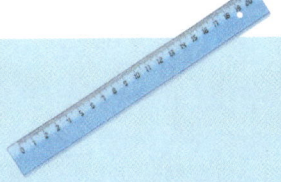  

Circle the correct tool.

2. Which tool measures how much milk is in a glass?

3. Which tool measures how long the book is?

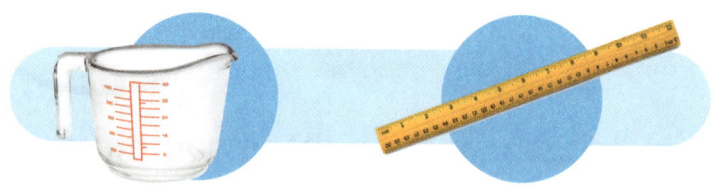 

4. Which tool measures around a tree trunk?

5. What can I use to see the caterpillar better?

A scientist uses tools to find answers that are correct. God made you with the ability to use tools to find answers.

Name _____

Recording Data

Why is it important to record what you observe?

Nate wants to be a scientist.
Nate likes to study insects.
He wonders whether he will see more
insects by the pond or in the garden.

He records, or writes down, what he observes.
Recording helps Nate remember what he saw.
It helps him find answers that make sense.

1. How many insects are by the pond? ____

2. How many insects are in the garden? ____

Nate observes 9 insects by the pond.
He observes 5 insects in the garden.

3. Color the bar graph to show the number of insects.

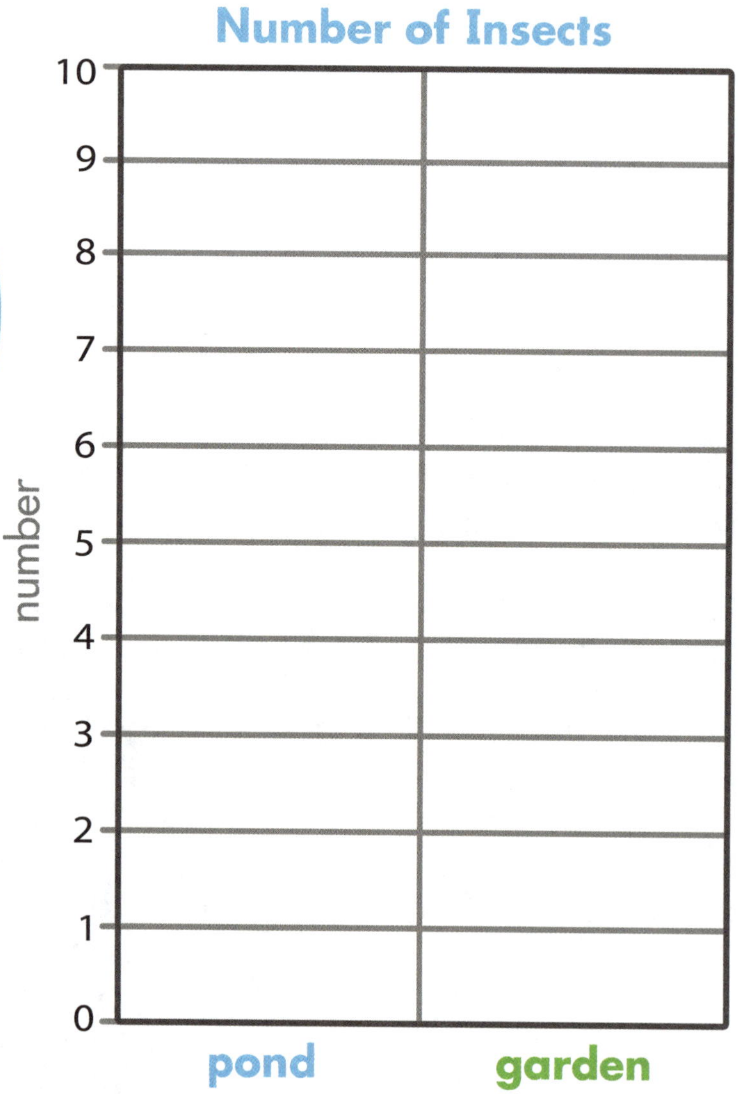

Number of Insects

pond   garden

Complete the exercises.

4. Which place has the most insects? _____

5. How many more insects are seen in the pond than in the garden? ____

6. What do you now know about insects and these places?

_____

_____

Name _____

Engineering

What does an engineer do?

God made people to create, or make, new things.
Engineers design and make new things.
Engineers do what God gave them the ability to do.

A man named George had a dog.
Burrs stuck to his dog's fur.
George got an idea.
He worked to design hook-and-loop tape.

Look at the items below.

1. What new thing can you make with each of these items?

_____

_____

God made people in His image.
He is the Creator.
People can create too.

Think about what you carry in a lunch box.

2. What would you like to carry? _____

3. Design a lunch box that can carry something new. Draw it here.

4. Write how your new lunch box solved a problem.

_____

_____

Name _____

Engineering Design Process

Engineers look for a solution to a problem. They follow steps to solve the problem.

Step 1 State the Problem

Think about a problem that needs to be solved.

I am hungry.

Step 2 Explore Ideas

Think of ways to solve the problem.

Should I buy food? Should I make food? Should I ask for food?

Step 3 Design a Plan

Choose a solution. Draw a model.

I will make a peanut butter sandwich.

You act like an engineer when you solve a problem.

Draw the three steps in the correct order.
- Set one slice on top of the other.
- Place bread on plate.
- Spread peanut butter.

© Science Grade 1

15

Step 4 Build and Test

Think about how to test the design.

I will spread peanut butter on bread. I will taste the sandwich.

Step 5 Analyze and Redesign

Discuss new ideas.

Think about how to design a better peanut butter sandwich.

Plan and draw your new design for a sandwich.

Step 6 Share Results

Show others how your solution solved the problem.

Share why this is a better solution.

Name _____

Chapter 1 Review

1. Match the sentence with the tool.

   Observe a rock more closely. •   •

   Measure the amount of tea. •   •

   Measure around a ball. •   •

   Find the length of a pencil. •   •

2. Fill in the circle next to the fair test.
○ Jess fills two glasses with the same amount of liquid.
○ Andy tosses a golf ball hard and a baseball softly.
○ Mel weighs a puppy before breakfast and another puppy after lunch.

Circle the word that best completes the sentence.

3. Scientists ask questions and look for **tools / answers** that make sense.

4. Look at the graph. Which ball bounced the highest? _____

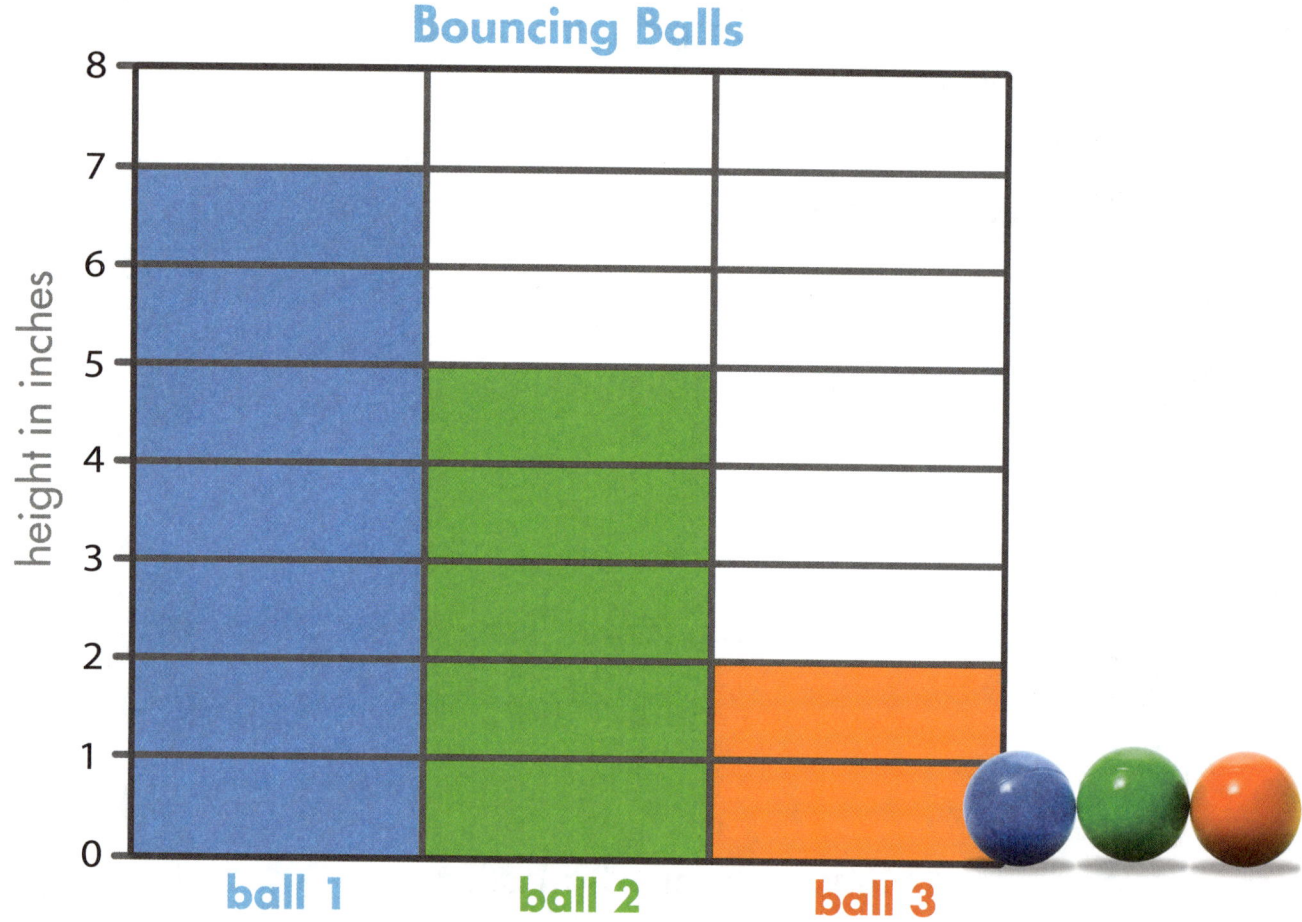

Fill in the circle next to the correct answer.

5. God designed you to __.
   ○ use science to ask questions and find answers that make sense
   ○ know all the answers to everything
   ○ let real scientists find all the answers for you

6. Which activity shows both a design and a test?
   ○ building a paper airplane and trying to fly it
   ○ running to the end of the road
   ○ choosing sticks instead of straws

7. Sasha makes a tool to help her get things that are out of reach, but it does not work. What should she do next?
   ○ break it and do something else
   ○ just use it because that is the only way to make it
   ○ try another way to make the tool and test it

Unit 2
Life Science

Look at the birds of the air; they do not sow or reap or store away in barns, and yet your heavenly Father feeds them. Are you not much more valuable than they?
Matthew 6:26

Name _____

Animals

Chapter 2

What kind of eye is this?

This is a false eye.
It makes the butterfly look bigger.

Why do you think God gave this butterfly colors that look like a big eye?

**Connect the dots to complete the picture.
Add color to match the wings.**

Name _____

Living and Nonliving

How do you know what is living or nonliving?

God made living and nonliving things.
Plants and animals are living things.
They need water, food, and air.
They grow and change.

Nonliving things are not alive.
They do not need water or food.
They do not grow.

1. Are you a living thing? _____

2. Circle three items living things need.

rocks   air   water   food

Classify means to put things that are alike in a group. You can classify living things and nonliving things.

3. Draw a blue rectangle around the living things. Draw a red rectangle around the nonliving things.

24

Name _____

Animals Have Needs

How does God care for an animal's needs?

Animals are living things.
Animals need water and food to live and grow.
God gives water for drinking and food for energy.
Animals eat plants and other animals.

1. What has God given these two animals? _____

2. What has God given these two animals? _____

Animals need places to live.
God designed animals so they can find
or make homes.

3. Make an X on each animal's home.

Living things also need air to breathe.
God put air on the earth.

4. In the correct word shapes, list the four things animals need.

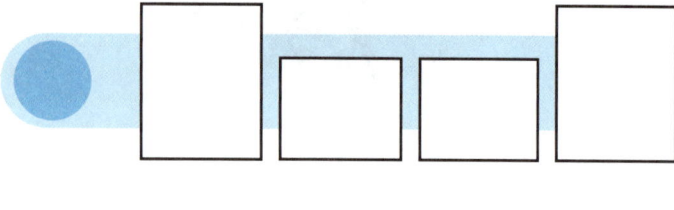

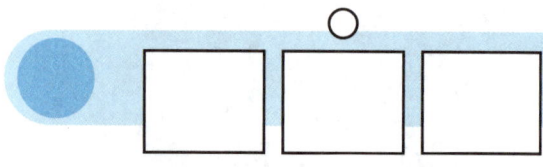

26

Name _____

2.3

Animals Have External Parts

How does God meet the needs of animals?
God designed animals to have parts that help meet their needs.

1. Circle one part on each animal that God made to meet a need.

butterfly

ibex

beaver

chameleon

27

God made many kinds of tongues.
He gave you a tongue to help you talk and eat.
Some animals have very long tongues.
These tongues help them get the food they need.

2. Circle which tongue is the longest.

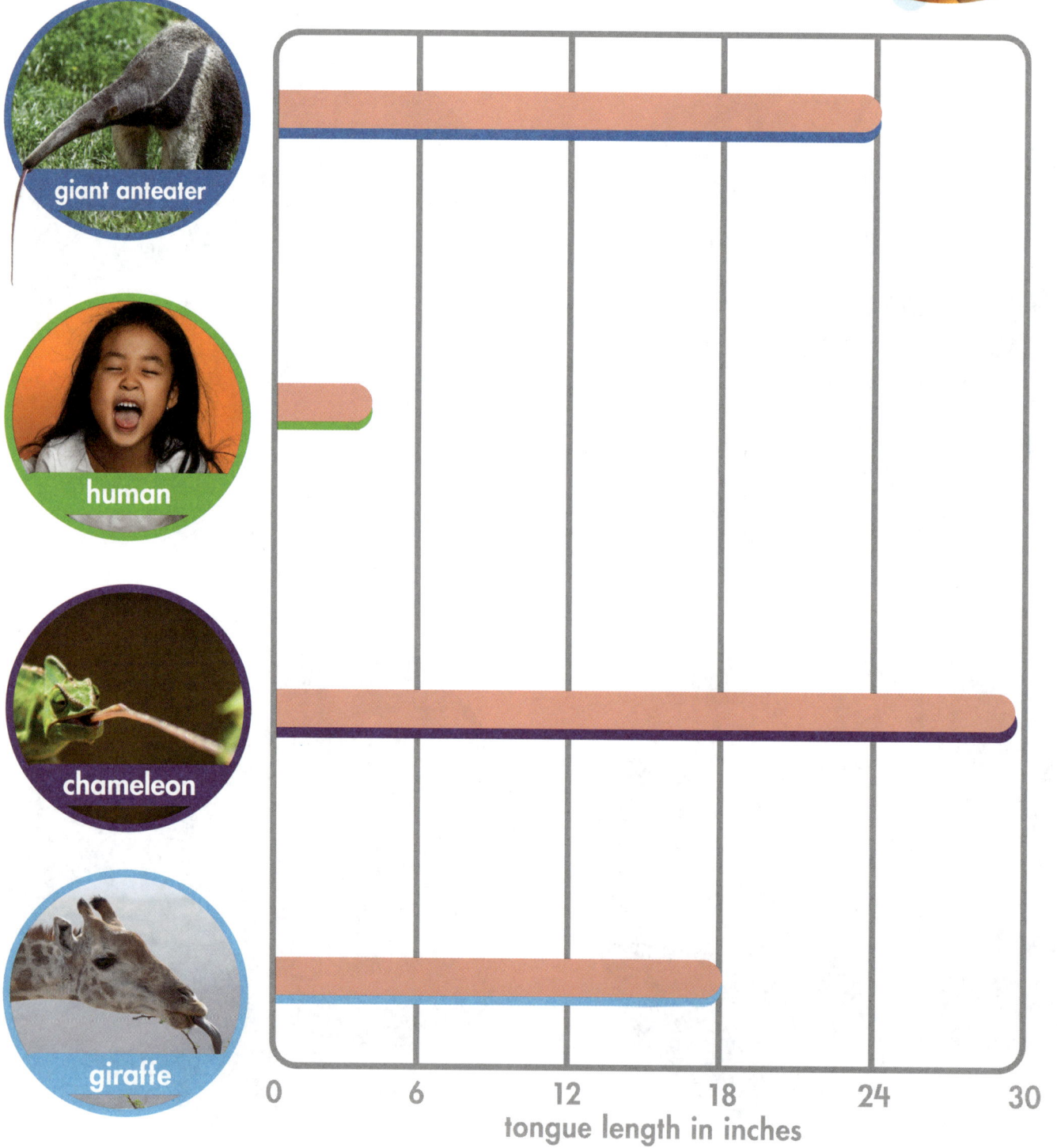

Name _____

2.4

Animals Can Be Grouped

What kinds of features did God give animals?

God gave animals different parts, like their body coverings.
Each animal has a body covering to help it live.

Fur, or hair, keeps mammals warm.
Dry scales keep reptiles cool.
Feathers help birds fly.
Scales protect fish in the water.
A hard shell protects insects.

Fill in the correct circle.

1. A beetle has a hard shell.
   A beetle is ___.
   ○ a reptile   ○ a mammal   ○ an insect

2. Fur keeps a deer warm.
   A deer is ___.
   ○ an insect   ○ a mammal   ○ a bird

3. Feathers help a finch fly.
   A finch is ___.
   ○ a bird   ○ a mammal   ○ a fish

God made many kinds of animals.
Mammals are one kind.
All <mark>mammals</mark> have hair and make milk to feed their babies.
Not all mammals live on land.

The elephant lives on land.
It has hair on its head, tail, and eyelashes.

The walrus hunts in the ocean and lives on land.
It has special hairs on its snout.

The dolphin lives in the ocean.
It has hairs along its chin and snout.

4. Draw one of these mammals and show its special hair. Write a sentence telling where it lives.

_____

_____

Name _____

2.5

Animals Babies and Their Parents

Are animal babies likes their parents?

A <mark>parent</mark> is an animal that has babies.
God made baby animals to be like their parents.
Many baby animals have the same shape as their parents.
The parents and their babies are the same kind of animal.

1. Draw a baby duck next to its parent.

31

Parents and babies sometimes look different. Parents and babies can have different colors.

2. Draw a line to match the baby to its parent.

- silvered leaf monkey
- emperor penguin  
- harp seal  
- tapir

Name _____

Body Coverings

Investigation 2.6

Step 1 — Ask a Question
Do body coverings keep animals warm?

Step 2 — Make a Prediction
I think ___ keeps animals warm.

○ fur
○ feathers
○ shells
○ no covering

Step 3 — Plan and Do a Fair Test

1. Place each body covering in a double-zipped bag. Make one empty double-zipped bag.
2. Tape a thermometer inside each bag. Record each temperature.
3. Set all 4 bags in a tub and fill the tub with ice.
4. Set a timer for 1 minute.
5. Record the temperature of each bag again.

Materials
- fake fur
- feathers
- eggshells
- 8 bags
- 4 thermometers
- tub
- ice
- timer

Step 4 — Record and Analyze Results

Complete the graphs.

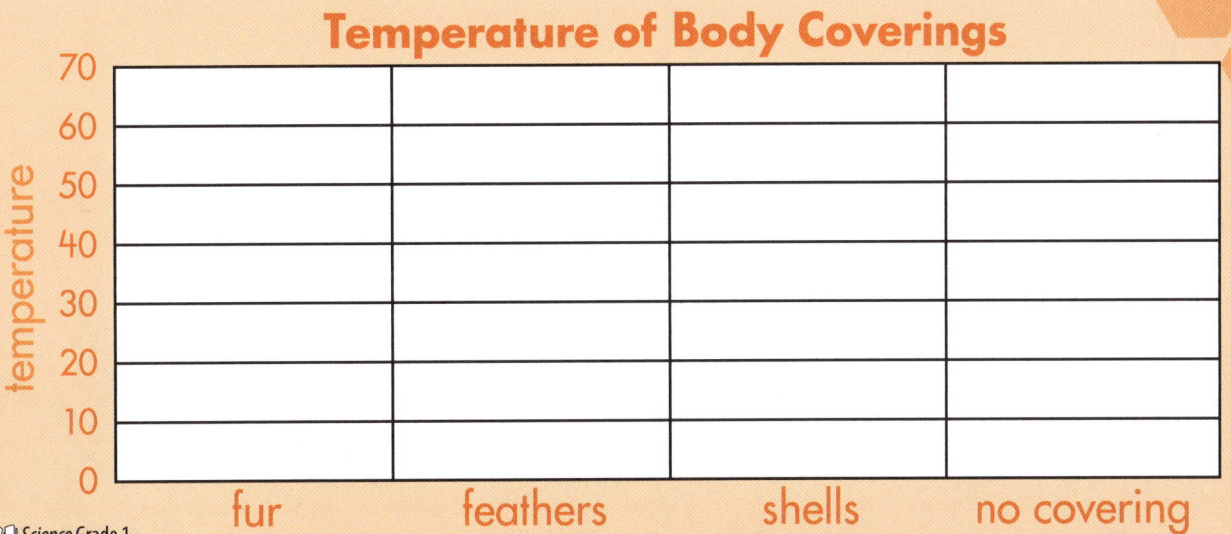

Temperature of Body Coverings

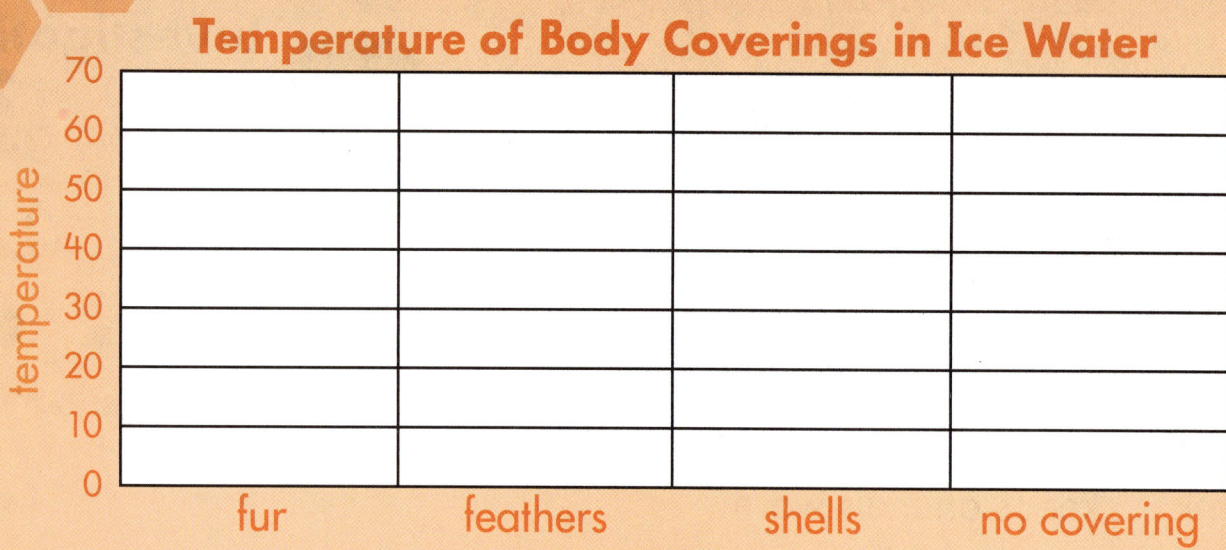

Step 5 Make a Claim

The body coverings **do / do not** keep animals warm.

Step 6 Give Evidence

After placing them in ice, the body coverings were at a **lower / higher** temperature than the empty bag.

Step 7 Share Results

Share your results with someone at home.

Think about why God made animals to have certain body coverings. Draw an animal that can live in the cold with its covering.

Name _____

Chapter 2 Review

Write the correct word on the line.

ANSWER BANK nonliving   living   mammal   covering   animals

1. A rock is a _____ thing.

2. A mouse is a _____ thing.

3. All _____ need food, water, air, and a home.

4. In the cold, an animal's body _____ can keep it warm.

5. A _____ has hair and makes milk for its babies.

6. Draw and color the correct body covering for each animal.

7. Make an **X** on the mammals.

Underline the correct answer.

8. What do reptiles and fish have in common?
They both have gills.
They both have scales.
They both have fins.

9. How are many baby animals like adult animals?
They are the same color.
They are the same size.
They are the same shape.

Name _____

Mammals

Chapter **3**

What made these prints?

These are dog prints. The shape of the print is the same as the shape of the paw.

The dog paw print has a heel.

1. How many toes does it have? ___
2. Draw four paw prints, or tracks, to get the adult dog to the puppy.

Name _____

3.1

What Mammals Are

How do you know it is a mammal?

Mammals have hair, or fur, and can make milk to feed their babies.

These cows have hair.
The mother cow makes milk to feed her baby.
The milk gives the calf, or baby cow, energy to grow.

1. What do mammals feed their babies? _____

Mammals need water, food, air, and shelter, or a place to live.

2. Look at the picture. Circle three things the cows need.

© Science Grade 1

39

Make a check mark next to the mammals.

9. What kind of body covering do mammals have? _____

Name _____

3.2

Mammals Have External Parts

Which parts help mammals live?

Mammals move to get food and shelter.
Which parts help them do this?

1. Fill in the circles for the parts that help the animal move.

○ webbed feet
○ fur
○ tail

○ claws
○ tail
○ fur

○ horns
○ legs
○ feet

2. Compare and contrast the parts of a beaver and a jackrabbit.

 beaver

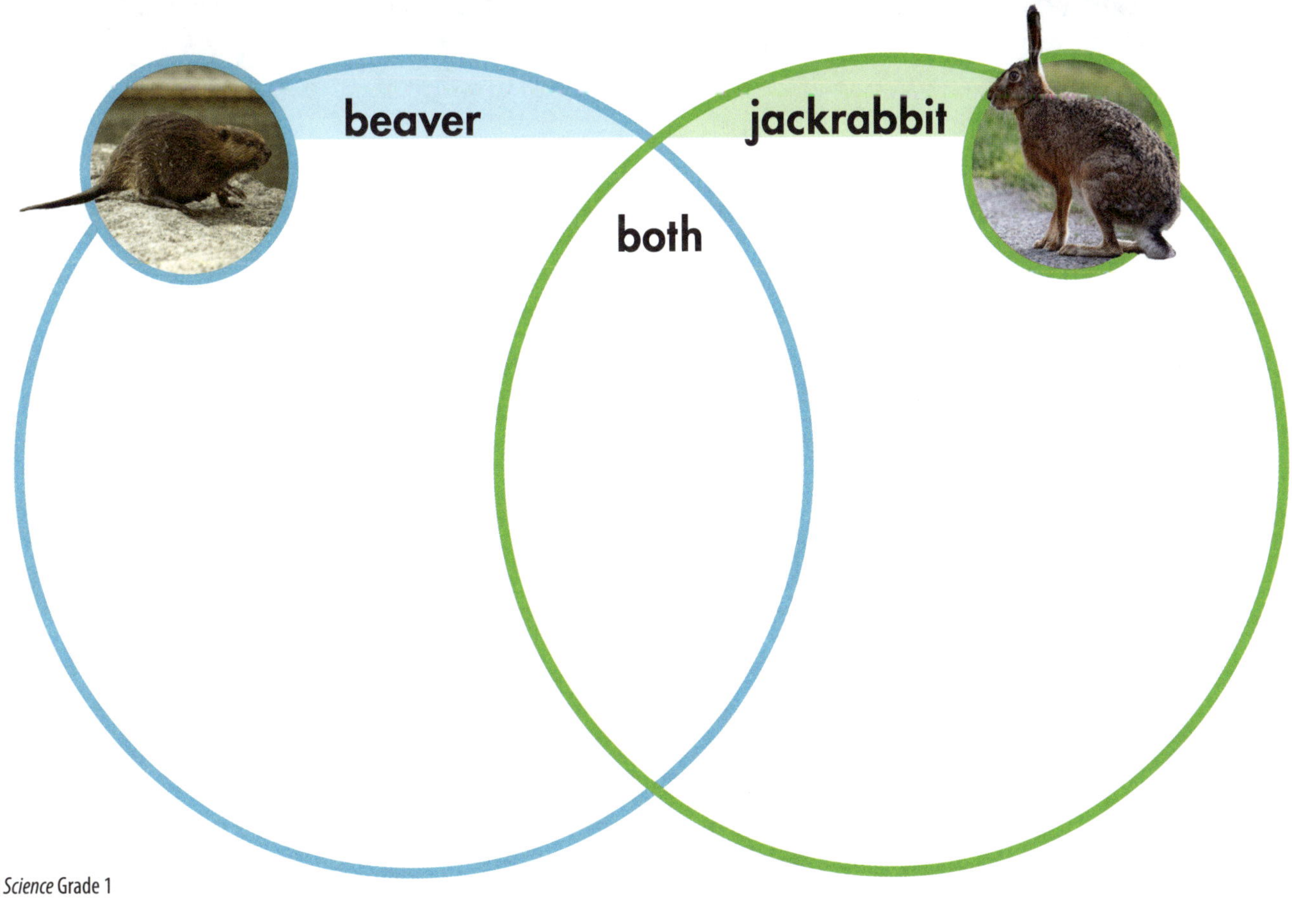

jackrabbit

both

41

Mammals move to get food and shelter.

Rob looked at the parts God gave some mammals to help them move.

Help him sort the mammals by the way they can move.

3. Fill in each blank with a word that tells how the group can move.
4. Write the names of the two mammals that go together under each group.
5. Discuss with a partner how you grouped the mammals.

Mammal Groups	
Group 1 has mammals that can _____.	Group 2 has mammals that can _____.

Name _____

Life Cycle of a Beaver

How do mammals care for their babies?

A **life cycle** is a series of stages in the lives of plants, animals, and people.

A mother beaver has a litter of kits.
After one year, the kits are bigger.
They are called yearlings.
A year later, they grow into juveniles.
They leave the lodge and find mates.
Each adult starts a new family.

1. Write the correct life-cycle stage on the line.

ANSWER BANK: yearling   juvenile   kit   adult

1 _____
2 _____
3 _____
4 _____

© Science Grade 1

43

God made mammals to care for their babies in different ways.

A mother beaver feeds milk to her kits for two weeks.
She teaches them how to find food and build a lodge.
After two years, the beavers leave their mother.

A mother orangutan feeds milk to her baby for five years.
She teaches her baby how to find food and build a nest.
After eight years, the orangutan leaves its mother.

Make a check mark in the square next to the correct answer.

2. Which mammal feeds its baby longer? ☐ beaver   ☐ orangutan

3. Which mammal stays with its mother a shorter amount of time?
☐ beaver   ☐ orangutan

4. Do all mammals care for their babies the same way? ☐ Yes   ☐ No

Name _____

Mammal Babies and Their Parents

How are mammal babies like their parents?

A mammal baby has the same body and face shape as its parent.

It has the same number of legs and similar hair.

Its feet are the same shape but not the same size.

Draw a line from the baby mammal to its parent.

1.  •

2.  •

3.  •

4. Help the baby mammals find their parents. Draw a line from each baby to its parent.

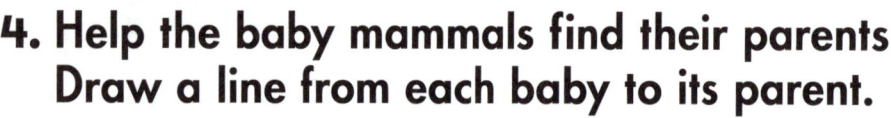

Name _____

3.5

Mammals of the Woodlands

What is in the environment?

An environment is all the living and nonliving things in an area. God puts living and nonliving things in the environment that animals need to stay alive.

The woodlands is an environment.
It has many trees and plants and plenty of water.

1. Draw two things in the woodlands that God gives the beaver.

Make an **X** on the animal that does not live in the woodlands.

2.

3.

4. Listen to your teacher to complete the page.

Name _____

Mammals Change Their Environments

Design
3.6

Imagine you are a beaver.
You want to build a home in a pond.
But first you need to make a pond.
Create a beaver dam to stop the river.

Materials
- twigs
- pebbles
- clay
- plastic box
- measuring cup
- water

Step 1 Ask a Question

Circle the best word.
The river flow must be **faster / stopped**.

Step 2 Explore Ideas

Talk about how you will use the materials to stop water from flowing.

Step 3 Design a Plan

How will you build a model of a beaver dam?
Draw a picture.

Step 4 Build and Test

Order the steps.

___ Watch what happens to the water when it gets to the dam.

___ Make a dam across the box.

___ Pour water into one end of the box.

Step 5 Analyze and Redesign

Complete the exercises.

1. Did any water get through the dam?
   ☐ Yes   ☐ No

2. How can you make the dam better?

   I will add more _____.

3. Make the changes and test again.

Step 6 Share Results

1. Talk with another group about how you solved the problem.
2. Share how beavers are part of God's design for their environment.

Name _____

Chapter 3 Review

Write the correct word from the Answer Bank on the line.

ANSWER BANK mammal   life cycle   milk   parent   legs

1. The environment has all the things a _____ needs to live.

2. A mammal baby has the same shape as its _____.

3. Mammals have hair and make _____ for their babies.

4. A beaver uses its _____ to swim and walk.

5. A _____ is a series of stages in the lives of animals.

Look at the picture. Complete the exercises.

6. Make an **X** on the things that meet the needs of the fox.

7. List one other thing a fox needs to live. _____

© Science Grade 1

51

8. Number the stages of a beaver's life cycle.

___   ___   ___   ___

Circle Yes or No.

9. All mammals care for their babies the same way.   Yes   No

10. A baby bear has the same foot shape as its parent.   Yes   No

11. God gives animals what they need in their environment.   Yes   No

12. Animals can change their environments.   Yes   No

Name _____

Fish and Birds

Chapter 4

Where do these grow?

These feathers grow on the top of a peacock's head. Together they are called a crest.

Complete the pattern of the body covering for the peacock.

Name _____

4.1

What Fish and Birds Are

Is it a fish or a bird?

A fish is an animal that lives in water.
A fish has fins, which are body parts that help it swim.
Scales cover the bodies of most fish and protect them.
Fish do not have legs.

1. Circle the name of the covering that protects the fish's body.

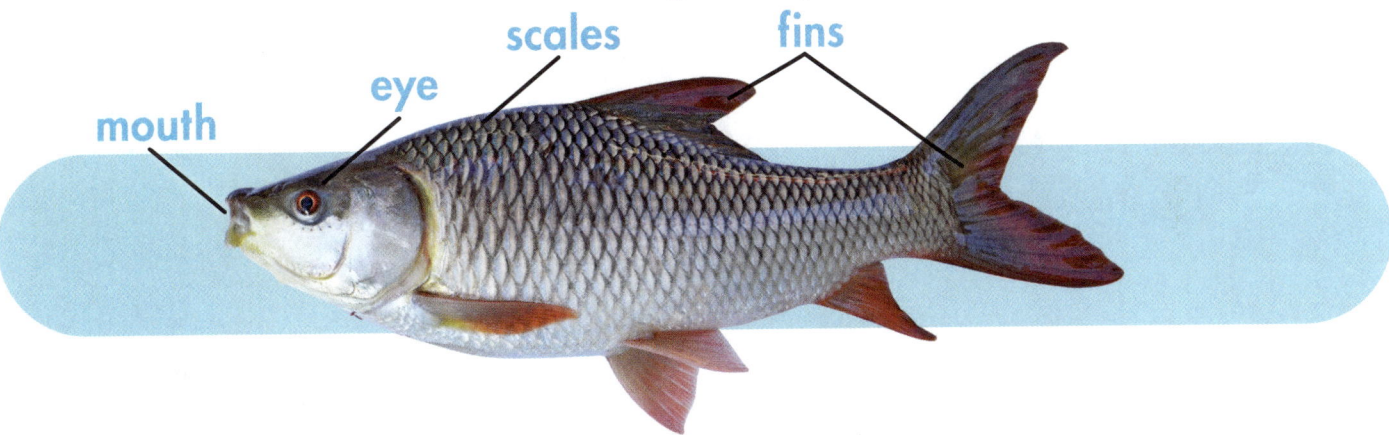

A bird is an animal that has a beak, the hard part of its mouth.
All birds have wings, but not all birds fly.
Feathers cover the body of a bird and protect it.
Birds have two legs.

2. Circle the name of the covering that protects the bird's body.

© Science Grade 1

55

3. Write the letter **F** next to the fish.

4. Write the letter **B** next to the birds.

5. What part of the animal tells you it is a fish? _____

6. What part of the animal tells you it is a bird? _____

Name _____

Fish Have External Parts

How does a fish move and breathe?

God gave fish all the parts they need to live and move.
Scales protect a fish from harm.
Fins and the shape of the fish help it swim.
Gills bring air from the water to the fish.

1. Connect the dots.
2. Add the three other parts the animal needs to live.

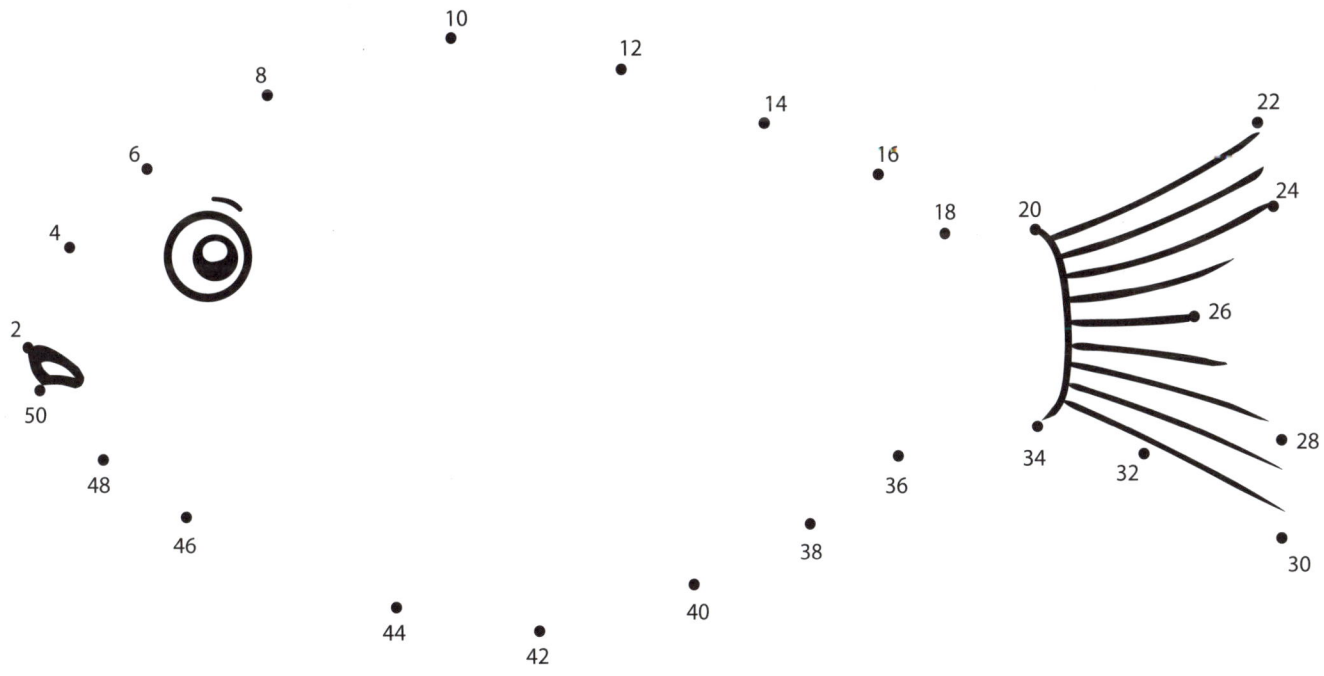

Why does a fish feel slimy?
Fish have mucus over their scales.
Mucus helps fish swim smoothly in water.
It also keeps germs from getting under the scales.

4.2

57

In God's design of fish, each part has a purpose.

3. Draw a line from each action to the part of the fish that does that action.

eat   breathe   see   move

4. Circle the human designs that have parts like a fish.

5. Draw a fish using these shapes.

Make sure the fish has all the parts it needs to live and grow.

Birds Have External Parts

The bird is the only kind of animal with feathers.
Feathers protect a bird's skin.
They keep a bird warm.

Feathers help a bird hide.

Feathers give a bird color.

Feathers help a bird fly.

Feathers are different sizes, shapes, and colors.

Write the correct word from the Answer Bank on the line.

| ANSWER BANK | warm | fly | animals | feathers |

1. Birds are the only _____ with feathers.

2. Feathers help a bird _____ and keep _____.

3. _____ help some birds hide.

God gave birds the kinds of feet they need for their environments.

4. Circle the word in each sentence that tells how the birds use their feet.

Clawed feet help the woodpecker climb.

Webbed feet help the mallard swim.

Clawed feet help the hawk grasp.

Birds eat and move in many kinds of environments. God designed a bird's beak to work for the kind of food it eats.

The woodpecker's pointed beak drills holes into trees to get insects.

The mallard's bill strains out water when it eats underwater.

The hawk's hooked beak tears its food.

5. Write the correct words from the Answer Bank on the line.

ANSWER BANK	tear swim drill climb grasp strain		
	Hawk	Woodpecker	Mallard
How does the bird use its feet?	_____	_____	_____
How does the bird use its beak?	_____	_____	_____

60

Name _____

4.4

How Fish and Birds Detect

Which parts help fish and birds find food and detect danger?

Fish and birds have eyes.
Birds have very good eyesight.
They have large eyes to help them see far away.
Birds use their eyes to find food and watch for danger.

Fish do not detect food or danger well with their eyes.
Fish have lateral lines to detect changes in the water.
Fish use their lateral lines to feel things around them.

1. Trace the lateral line on the fish.

Fish and birds have ears.
But it is not easy to see the ears.
They are hidden under the scales or feathers.

2. Circle the eye and draw an ear on the fish and the bird.

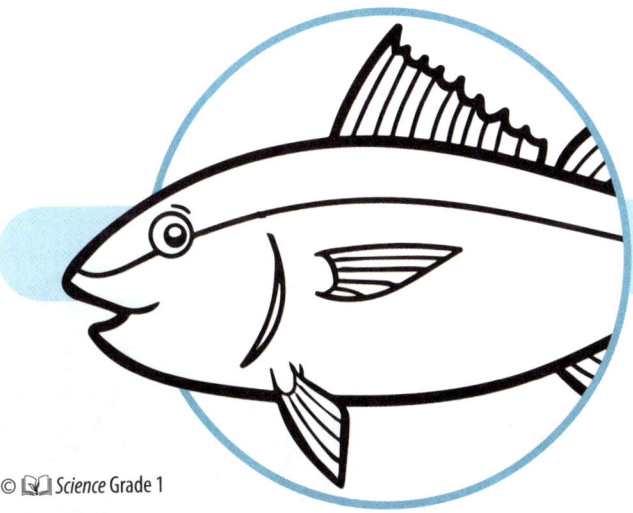

Birds and fish can smell and taste.
They use nostrils to smell.

Birds use their tongues to taste, but fish use their whole heads.
Catfish also have barbels that taste the bottom of a lake to find food.

3. Circle the parts that help the catfish find food.

God gave animals ways to protect themselves when they detect danger.
The puffer fish puffs up when it detects danger.
Its larger size makes it look too big for other fish to eat.

4. Circle the puffer fish that detects danger.

God made some fish to live in groups called schools.
Schools protect fish from danger.
If a shark comes, fish move together to look like one big fish.
The school may circle around the shark.
This moving makes it hard for the shark to get fish.

Name _____

4.5

Life Cycles of Fish and Birds

How do fish and birds grow and care for their young?

Fish and birds have life cycles like other animals.

Animals are called <mark>young</mark> when they are in an early stage of life.
First, the young hatch from eggs.
Then, the young grow to be adults.

1. Draw the rest of the life cycle for the fish and the bird.

2. Look at the life cycles. What do fish and birds have in common?

_____

3. What are in the eggs? _____

© Science Grade 1   63

Most fish lay eggs.
Their young hatch from the eggs.
Many eggs float in the ocean and are eaten by other fish.
Some fish hide their eggs.
Some fish build bubble nests for their eggs.
They protect their eggs from predators.

All birds lay eggs.
The eggs hatch.
Young birds grow into adults.
Most birds build nests for their eggs.
Birds care for their young and protect them.

Write the correct words from the Answer Bank on the lines.

ANSWER BANK
protect
adult
nests
eggs

4. Some fish hatch from _____ and then grow up to become _____ fish.

5. Birds build _____ for their eggs to _____ their young.

Name _____

Investigation 4.6

Camouflage

Step 1 — Ask a Question

Does the color or pattern of a fish's scales help keep the fish safe from a predator?

Materials
- an environment
- 15 "fish" that match the environment
- 15 "fish" that do not match the environment
- timer
- paper cup

Step 2 — Make a Prediction

Circle your prediction.
A fish is safer when the color or pattern of it scales is **like / not like** its environment.

Step 3 — Plan and Do a Fair Test

Make a model and color the pictures below to match your model.

1. Choose an environment.
2. Add 15 fish that match the environment to be Fish A.
3. Add 15 fish that do not match the environment to be Fish B.
4. Color the pictures to look like the environment, Fish A, and Fish B.
5. Give a predator 10 seconds to grab fish one at a time and to place them in a cup.
6. Record how many of each fish the predator grabbed.

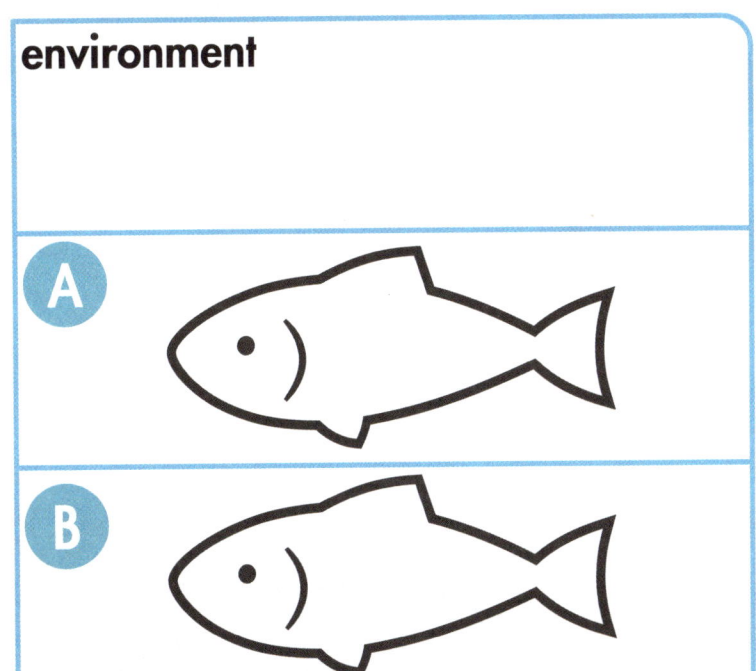

Step 4 — Record and Analyze Results
Graph your results.

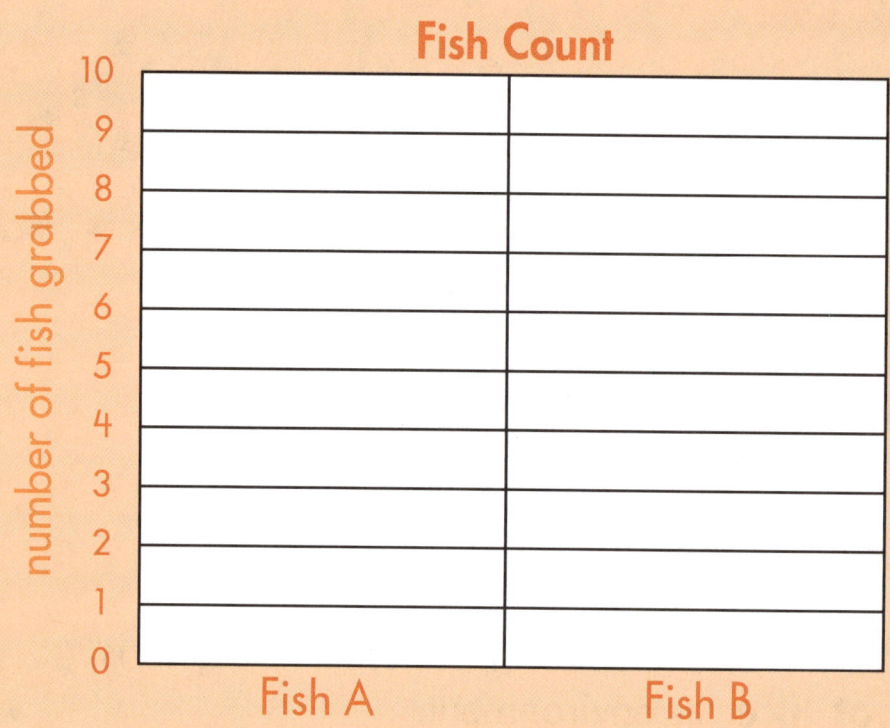

Step 5 — Make a Claim
Circle Yes or No.
Does the color or pattern of a fish's scales help keep it safe?   Yes   No

Step 6 — Give Evidence
How do you know your claim is true?

Step 7 — Share Results
Circle Yes or No.
Share your claim and evidence with another group.
Is their claim the same as yours?   Yes   No

Name _____

4.7

Chapter 4 Review

Use the Answer Bank to make a true statement.

ANSWER BANK   fins   beak   young   predator

1. A fish uses its _____ to swim in the water.

2. A _____ is an animal that eats other animals.

3. Parent birds take care of their _____.

4. The shape of a _____ helps a bird eat its food.

5. Make an **X** on the picture of a parent caring for its young.

6. Order the stages of the bird life cycle.

7. Use the Answer Bank to label the body coverings and the parts.

ANSWER BANK: gill cover   mouth   beak   legs   scales   fins   wings   feathers

Write Yes or No.

8. God designed a green and brown fish to hide in plants. _____

9. Fish and birds have legs. _____

10. Baby birds chirp so their parents will feed them. _____

11. The shape of a duck's feet helps it climb trees. _____

Name _____

Plants

Chapter 5

What kind of flower is this?

This is not a flower.
It is an insect that looks like a flower.

God provides many animals with camouflage to blend in with their environments.

**Color the plant and the insect.
Give the insect camouflage so it blends in with the plant.**

Name _____

Plants Have Parts

What parts does a plant have?

Plants are living things.
They have parts to help them live and grow.

A <mark>leaf</mark> makes food.
It uses energy from the sun.

The <mark>stem</mark> holds up the plant.
It carries water to the leaves.

A <mark>root</mark> holds the plant in the ground.
Roots take in water.

**Look at the plant picture.
Complete the exercises.**

1. Which part makes food? _____
2. Add two of them to the plant.
3. Highlight the part that takes in water.

God gave plants roots, stems, and leaves.
These parts help plants live and grow.
God made plants in many shapes and sizes.

4. Make an **X** on the part that holds each plant in the ground.

Circle the words for the correct parts.

5. Which parts of the plant can you see?
   roots   stem   leaves

6. Which part of the plant can you not see?
   roots   stem   leaves

Name _____

Life Cycle of a Plant

Are seeds part of the plant life cycle?

Most plants begin as seeds.
The seed is the first stage in the plant life cycle.

God designed each kind of plant to have its own kind of seed.
The seeds are as different as the plants.

Complete the exercises.

1. Order the stages of the plant life cycle.

2. Draw arrows to connect the dots.

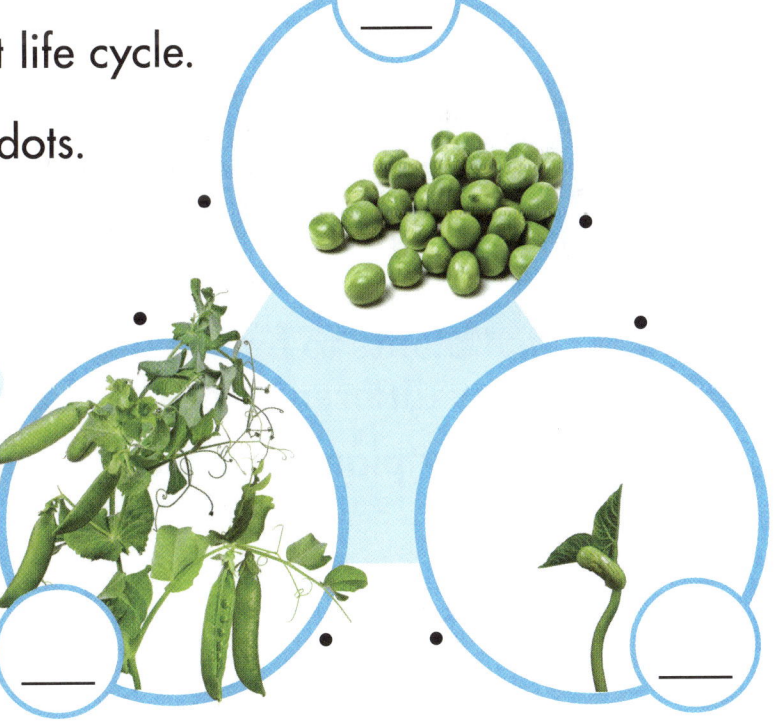

3. Circle the leaves in every stage they can be seen.

4. Where do seeds come from? _____

Plants are very much like their parents.
Bean seeds grow bean plants that look like the parent plant.

Make an X on the plants that do not come from the same seed as the first plant.

5.

6.

God designed plants to have a life cycle.
He made many different kinds of plants.

7. Create a new plant and draw each stage of its life cycle.

Name _____

5.3

What Plants Need

Where do plants grow?

Plants need air, water, sun, food, and places to live and grow.
Many plants grow in soil, or dirt.
The soil is not the same for all plants.

Some plants need more water than other plants.
Plants that need lots of water may have dark green leaves.
Plants that need a little water may have thick leaves.

1. Fill in the circle next to the sentences that are true.

  ○ All plants need places to live and grow.
  ○ All plants need the same kind of soil.
  ○ All plants need water.

God has placed plants in environments that meet their needs. Plants that need little water can live and grow in a desert. Plants that need more water can live in the woodlands.

Yucca plants grow in very dry environments.

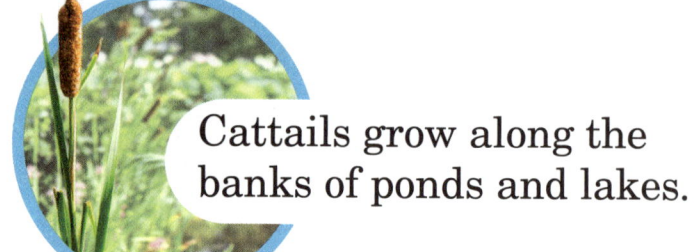

Cattails grow along the banks of ponds and lakes.

Complete the exercises.

2. Circle the environment that gets less water.

3. Would a yucca or cattail live in the environment you circled? Draw the plant that needs less water in that environment.

4. Write the plant needs that are missing from the list.

- air
- _____
- _____
- _____
- places to live and grow

76

Name _____

Plants Are Useful

How do animals use plants?

Energy comes from the sun.
Plants use the energy to make food.
Some animals eat plants to get energy.
Some animals eat other animals that eat plants to get energy.

1. **Draw arrows to show how energy moves from the sun to the animal.**

Kangaroos eat grass, leaves, and insects.

2. **Draw a line to what gives the kangaroo energy to hop.**

Eating food gives people energy.

3. **Circle the foods that come from plants and give you energy.**

God places plants in many environments.
Animals use the plants for food and shelter.

4. Make an X on the animal not using a plant.

5. Draw an animal in its environment using a plant to meet its need. Label the animal and environment.

ANSWER BANK

mammal	bird	food	shelter
rain forest	fish	desert	woodland

6. In your picture, which need is being met? _____

Name _____

5.5

Plants Protect Themselves

How do plants protect themselves?

Many plants have thorns or prickles to protect them.

The plant hopper is a small insect.
It can fit between the prickles.

hawthorn plant

Spines on a cactus keep many animals away.
Some animals can still get water from the cactus.

1. Circle the plant that protects itself.

God protects some plants in ways you cannot see.
Poison ivy gives a bad rash when it is touched.
Virginia creeper is harmless.
But poison ivy and Virginia creeper look very much alike.

2. Compare the two plants and fill in the chart.

	Poison Ivy	Virginia Creeper
Shape of Leaves		
Number of Leaves in Group		
Color of Leaves		

3. How do you know if a plant is poison ivy or Virginia creeper?

_____

_____

Name _____

Seed Growth

Investigation 5.6

Step 1 Ask a Question

Which will grow longer, the _____

or the _____?

Step 2 Make a Prediction

After 4 days, I think the _____ will grow longer.
Draw a picture of your prediction.

Materials
- ziplock bag
- moist paper towel
- 3 radish seeds
- ruler

Step 3 Plan and Do a Fair Test

Check off each part of the test as you finish it.

☐ **1.** Place the wet paper towel in the ziplock bag.

☐ **2.** Add 3 radish seeds to the bag.

☐ **3.** Gently press the bag closed and seal it about halfway.

☐ **4.** Hang up the bag and wait.

Step 4 Record and Analyze Results

Wait 4 days.

1. Circle the tool that can measure the length of the plant part.

2. Measure the stem and root.

stem length _____   root length _____

Step 5 Make a Claim

The _____ is longer.

Step 6 Give Evidence

How do you know your claim is true?

_____

_____

Step 7 Share Results

Share your claim and evidence with another group.
Complete the exercises.

1. How are the results the same as the other group's results?

_____

_____

2. How are the results different from the other group's results?

_____

_____

3. How did God design a plant to start its life? Circle one.

   more root   more stem

Name _____

5.7

Chapter 5 Review

1. **Write the letter that describes the plant part.**
   A. It makes food for a plant.
   B. It holds the plant in the ground.
      It takes in water from the soil.
   C. It holds up a plant.
      It carries water to other parts of the plant.

Write the correct word on the line.

ANSWER BANK: shelter   needs   energy   defense

2. God gave some plants spines, thorns, or prickles for _____.

3. Animals use plants for food and _____.

4. Animals that do not eat plants get _____ by eating animals that eat plants.

5. God places plants in an environment where their _____ can be met.

6. Draw the missing stage in a plant's life cycle.

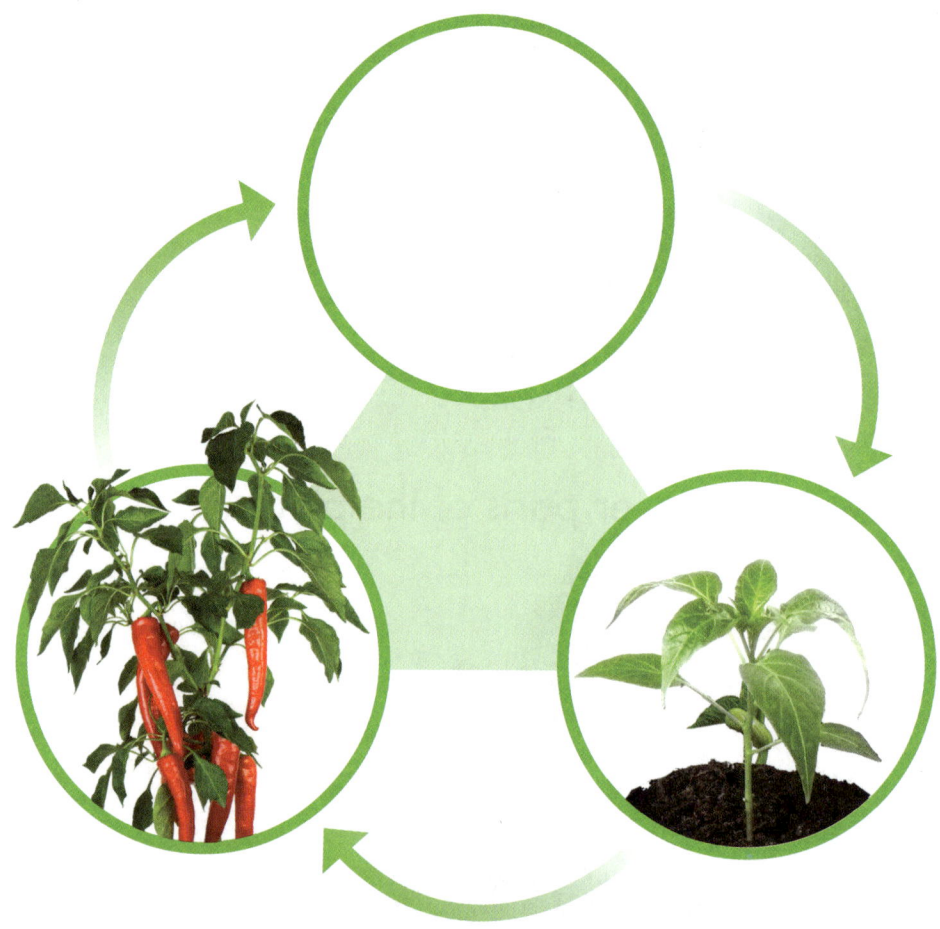

Circle the correct word.

7. The leaves on a **seed / young plant** are similar to the leaves on the same kind of adult plant.

8. An owl might use a tree for **shelter / soil**.

9. A plant that looks brown and dry is **healthy / unhealthy**.

10. Plants need a place to live and grow. List four other things they need.

_____ _____ _____ _____

11. Why are there no tropical plants in a desert?

_____

_____

Name _____

Career
Veterinarian

A veterinarian, or vet, is a doctor for animals.
Vets care for sick animals.
They go to college for seven to nine years.
People who are vets help all kinds of animals.

Vets help big animals like cows and horses.

Vets help small animals and pets like dogs and cats.

Some vets care for zoo animals like tigers and apes.

Unit Connect

Complete the exercises.

1. What kind of animals are the vets caring for on this page?
Circle the correct answer.   birds   mammals   fish

2. Draw a picture of a vet with an animal that is not a mammal.

Biography
James Herriot

James Herriot was a vet.
People took their sick pets to him.
He drove to farms to care for cows and sheep.
He fixed broken bones and cleaned wounds.
James wrote books about his life as a vet.
In one story, he saved a tiny kitten from the cold.
In another story, James helped a hurt horse.

Write the correct answer on the line.

1. What kind of job did James Herriot have? _____

2. Where did he drive to care for cows and sheep? _____

Draw a line under the correct word.

3. A vet is like a **farmer / doctor**.

Name _____

Glass Toy Case

Lee has a special toy for show and tell.
He wants to bring it to school, but he is afraid.
His toy is made of glass.
It could break on the way to school.
How can Lee get his toy to school without breaking it?

Step 1 State the Problem
Think about the problem you need to solve.

1. I need to get Lee's **backpack / toy** to school without it **breaking / getting lost**.

2. Circle all the actions that could cause Lee's toy to break.
   falling on the floor
   bumping inside his backpack
   sitting on a shelf
   being held carefully
   bumping into someone

Step 2 Explore Ideas

Think about what you know about animals.

What parts did God give animals to protect their babies?

1. Circle the babies that are protected.

God made kangaroos to have pouches.
The parent kangaroo can hop with her baby tucked in her pouch.

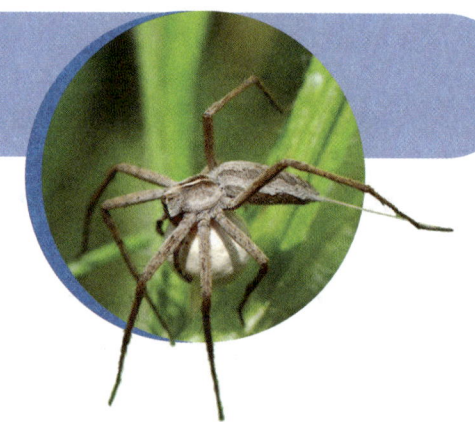

God designed spiders to make silk.
The silk wraps the eggs.
This parent spider can walk and climb with her eggs wrapped up safely.

2. Draw a rectangle around the human designs that protect like a pouch. Circle the designs that protect like a spider's silk wrap.

Name _____

Glass Toy Case

Step 3 Design a Plan

Use an egg as a model of Lee's toy.
Design a case that will carry an egg through three tests.
Your case should keep the egg from breaking.

1. My case will be designed like a **pouch** / **wrap**.

2. List materials you will use.

   _____

   _____

   _____

3. What is one material you will not use? Why?

   _____

   _____

   _____

4. Draw a picture of your design.
   Label the materials.
   Draw an arrow to show where the egg will go.

5. What part of your design might be hard to make?

_____

_____

_____

Name _____

Glass Toy Case

Step 4 Build and Test

Use your drawing to build your case.

1. Draw the case after you have built it.

2. Was your design easy to make? _____

3. What was the hardest part to make? _____

Test your case.

Place your case inside a ziplock bag.
Place the bag inside a backpack.
Do the three tests your teacher gives you.

Record your results.

Tests	Did your case work?	
Test 1	☐ Yes	☐ No
Test 2	☐ Yes	☐ No
Test 3	☐ Yes	☐ No

Name _____

Glass Toy Case

Step 5 Analyze and Redesign

Analyze the results.

1. Did you solve the problem? ☐ Yes ☐ No

2. What worked the best?

   _____

3. What needs to be fixed?

   _____

4. How could you make your design better?

   _____

Redesign.

5. Draw a picture of your new plan.
   Label the materials you will use.
   Draw an arrow to show where the egg will go.

Step 6 Share Results

Tell others what you learned from your design.

1. Use the chart to compare five cases.

Case	Did it work?	How was it designed?	What main materials were used?
1	☐ Yes ☐ No	☐ pouch ☐ wrap	
2	☐ Yes ☐ No	☐ pouch ☐ wrap	
3	☐ Yes ☐ No	☐ pouch ☐ wrap	
4	☐ Yes ☐ No	☐ pouch ☐ wrap	
5	☐ Yes ☐ No	☐ pouch ☐ wrap	

2. Which kind of design works the best? Why do you think that?

_____

_____

_____

Unit 3
Physical Science

The voice of the Lord strikes
with flashes of lightning.
... And in His temple all cry, "Glory!"
Psalm 29:7, 9

Name _____

Movement and Sound

Chapter 6

What is a woofer?

A woofer might sound like it is a part of a dog. But it is the part of a speaker that makes low sounds.

Trace the path the dog takes to get to the music.

Name _____

What Makes Things Move

How do living and nonliving things move?

Living things can move on their own.
They use their own energy.
Nonliving things cannot move on their own.
They need to be moved by something else.

Circle the correct answer.

1. Bees move on their own because they are **living / nonliving**.

2. Circle the thing that cannot move on its own.

3. How many children move each way?
Make a tally mark for each child.

climb		jump	

God gives living things energy from the sun.
Plants use energy to grow and to turn to the sun.
Animals use energy to find homes and food.

People use energy to sing and worship God.
They use energy to run and play.
People also use energy to move nonliving things.

**Look at the picture below.
Complete the exercises.**

4. Circle the children who are moving their own bodies.

5. Draw a rectangle around the child moving a nonliving thing.

Name _____

Ways Things Move

What are different ways things move?

Things move in different ways.

Pip the rabbit moves in different ways.

Trace the path using the colors shown below.

1. 🔴 First, Pip goes straight between the lettuce plants.
2. 🟠 Then, Pip moves round and round the baskets.
3. 🔵 Next, Pip zigzags from flower box to flower box.
4. 🔴 Finally, Pip runs straight to get out of the garden.

101

 A <mark>push</mark> moves things away from you.

 A <mark>pull</mark> brings things toward you.

Look at the pictures and circle the correct answers.

5. Is it a push or a pull that moves the swing?
    push   pull

6. What must one team do to win?
    push harder   pull harder

A moving thing can bump another thing and make it move too.

Look at the picture to complete the exercises.

7. Which thing will make the ball move?
    flag   girl's shoe   golf club

8. Draw a red arrow from the golf club to the ball.

9. Draw a blue arrow from the ball to the hole.

10. Which moves farther?
    golf club   golf ball

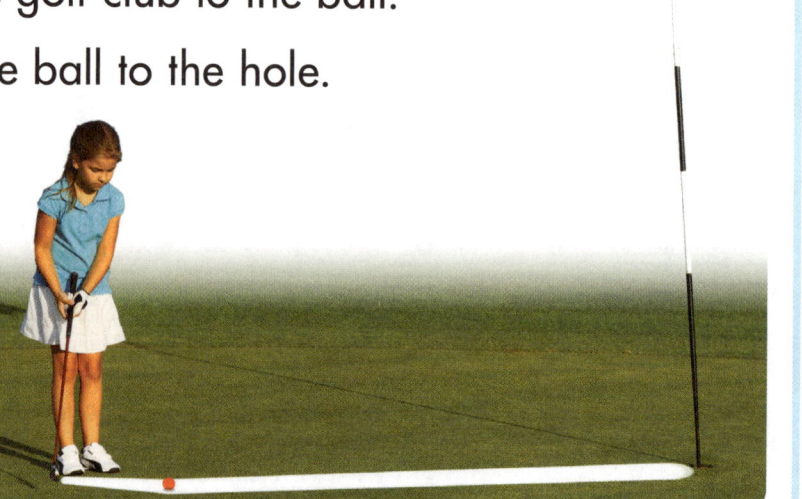

Name _____

Investigation 6.3

Changing Motion

Step 1 — Ask a Question
How do you move a ball through a golf course?

Step 2 — Make a Prediction
Circle your predictions.

1. A **smaller** / **bigger** push will move the ball farther.

2. When the ball hits something, it will change **direction** / **size**.

Step 3 — Plan and Do a Fair Test

1. Design a mini golf course. Draw it in the rectangle.

2. Write the steps to test how hard you need to hit the ball.

_____

_____

3. Test your course and make any changes needed to the design.

Step 4 — Record and Analyze Results

1. Draw a picture of the golf course that has the changes you made.
2. Draw arrows to show the path of the ball each time you hit it. Use red for a bigger push and blue for a smaller push.

3. How many pushes did it take to get through the golf course? ___
4. I used **smaller** / **bigger** pushes when the ball had to go farther.
5. When the ball hit something, it changed **direction** / **size**.

Step 5 — Make a Claim

Different / **The same** kinds of pushes move the ball through the golf course.

Step 6 — Give Evidence

I used both bigger and smaller **pushes** / **pulls** to move the ball.

Step 7 — Share Results

What did you learn about the ways God designed things to move?

_____

_____

Name _____

6.4

Making Sound

What makes sound?

When something moves back and forth very fast, it <mark>vibrates</mark>.
Sound is made when something vibrates.
Adding a push or a pull can make something vibrate.

**Write push or pull to show what makes the sound.
Circle the part that vibrates.**

1. drum _____

2. xylophone _____

3. guitar _____

4. harp _____

5. Complete the chart.

ANSWER BANK: vibrates   sound   strings

cause	effect
A guitar is strummed.	The _____ vibrate.
A rubber band is plucked.	It _____ back and forth.
A tuning fork vibrates.	It makes a _____.

© Science Grade 1

105

Sound is important to life.
God made sound to help people know things.

Match the reason for the sound to the item that makes the sound.

6. It is time to start the game.

7. Someone wants to come inside.

8. A friend wants to talk to you.

9. We celebrate a holiday.

10. There is a fire.

Sounds tell you how people feel.
The children giggle.

11. How do they feel?

_____

106

Name _____

6.5

Nature of Sound

How are sounds different?

Music plays.
Children clap.
A friend whispers in an ear.
Sounds are everywhere.
Some sounds are loud.
Other sounds are soft and
may be hard to hear.

1. Circle the picture of people making soft sounds.

2. Order the pictures from softest to loudest.

____ whisper   ____ train noise   ____ clap

Different things make different sounds.
Some sounds are high like the sound of a flute.
Some sounds are low like the sound of a tuba.
A mix of high and low sounds can make music.

Small bells make high sounds.
Big bells make low sounds.

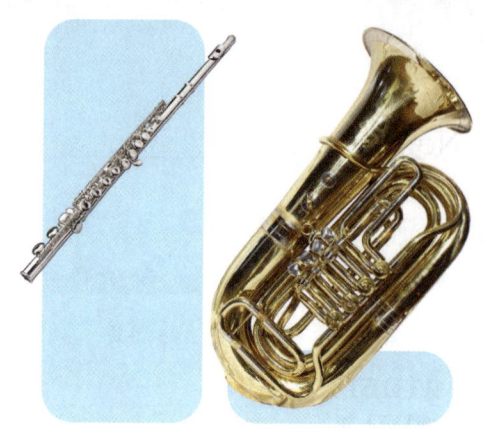

Look at the bells and complete the exercises.

3. Draw a bell that has a lower sound than the other bells.

4. The bell I drew is **bigger / smaller** than the other bells.

God made animals that have different calls.
The mother cat calls her baby and hears its cry.
She knows the kitten's cry and finds it.

5. Draw the kitten in the correct tree.

Name _____

Investigation 6.6

Testing Sound

Step 1 Ask a Question

What sounds can make sugar vibrate?

Step 2 Make a Prediction

Write the sounds you will test.

Materials
- prepared bowls
- sugar
- 4 sounds

Sound	Do you think the sugar will vibrate?
1. _____	Yes No
2. _____	Yes No
3. _____	Yes No
4. _____	Yes No

Step 3 Plan and Do a Fair Test

Follow the directions to test the sounds.

1. Choose four sounds you will test.
2. Make each sound as close to the sugar as you can.
3. Observe whether the sugar vibrates.
4. Record your results.

Answer the question.

5. How will you know whether a sound makes sugar vibrate?

I will see _____.

Step 4 Record and Analyze Results

Complete the chart.

Sound	Did the sugar vibrate?
1. _____	Yes No
2. _____	Yes No
3. _____	Yes No
4. _____	Yes No

Step 5 Make a Claim

Circle the correct word.
Sound from one object **can / cannot** make another object vibrate.

Step 6 Give Evidence

Finish the sentence.
I saw the sugar vibrate when

_____.

Step 7 Share Results

Discuss with other groups which sounds make sugar vibrate.

Fill in the circle for the correct answer.
What does this investigation prove about God's design of sound?
○ God made sound to not have energy.
○ God made sound to be able to vibrate things.
○ God made all things sound the same.

110

Name _____

Chapter 6 Review

Fill in the circle for the correct answer.
How is the child moving?

1.
   ○ by her own energy
   ○ by a push
   ○ by a pull

2.
   ○ by his own energy
   ○ by a push
   ○ by a pull

3. How did God design living things to get energy?
   ○ They get energy from what they eat.
   ○ They get energy from sound.
   ○ They get energy by moving.

Circle the correct answer.

4. Which ball needs the biggest push to get to the hole?

5. How will the ball need to move?
   zigzag   round and round   straight line

111

6. Match the cause with its effect.

I beat a drum. •   • The drum gets louder.

I beat the drum harder. •   • The drum vibrates and makes a sound.

7. Color the part of the drum that vibrates.

8. Color the pipe that makes the lowest sound.

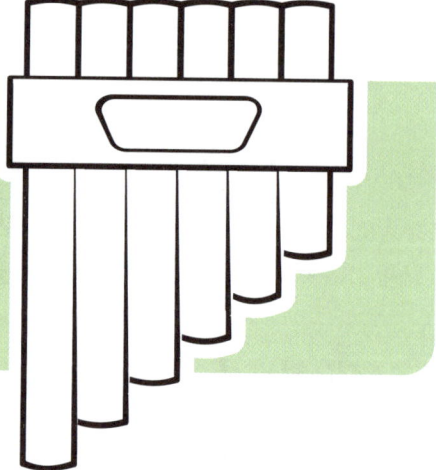

9. Circle the pictures that show people using sound to tell something.

Name _____

Light

Chapter **7**

What made these shadows?

Horses made these shadows. When light hits a living or nonliving thing, it creates a shadow.

God created the sun. Its light caused these shadows. What other kinds of light could cause shadows?

Draw what your shadow looks like.

Name _____

7.1

Seeing Light

What is light?

God created light. Light is a kind of energy that makes you able to see things. Most of the earth's light comes from the sun.

Some things make their own light. Others do not. Things that do not make their own light must have light shining on them to be seen.

1. Draw an **X** over the things that make their own light.

115

Some light is made by God. Some light is made by people.

2. Write **G** under items that are made by God.
Write **H** under items that are made by humans.

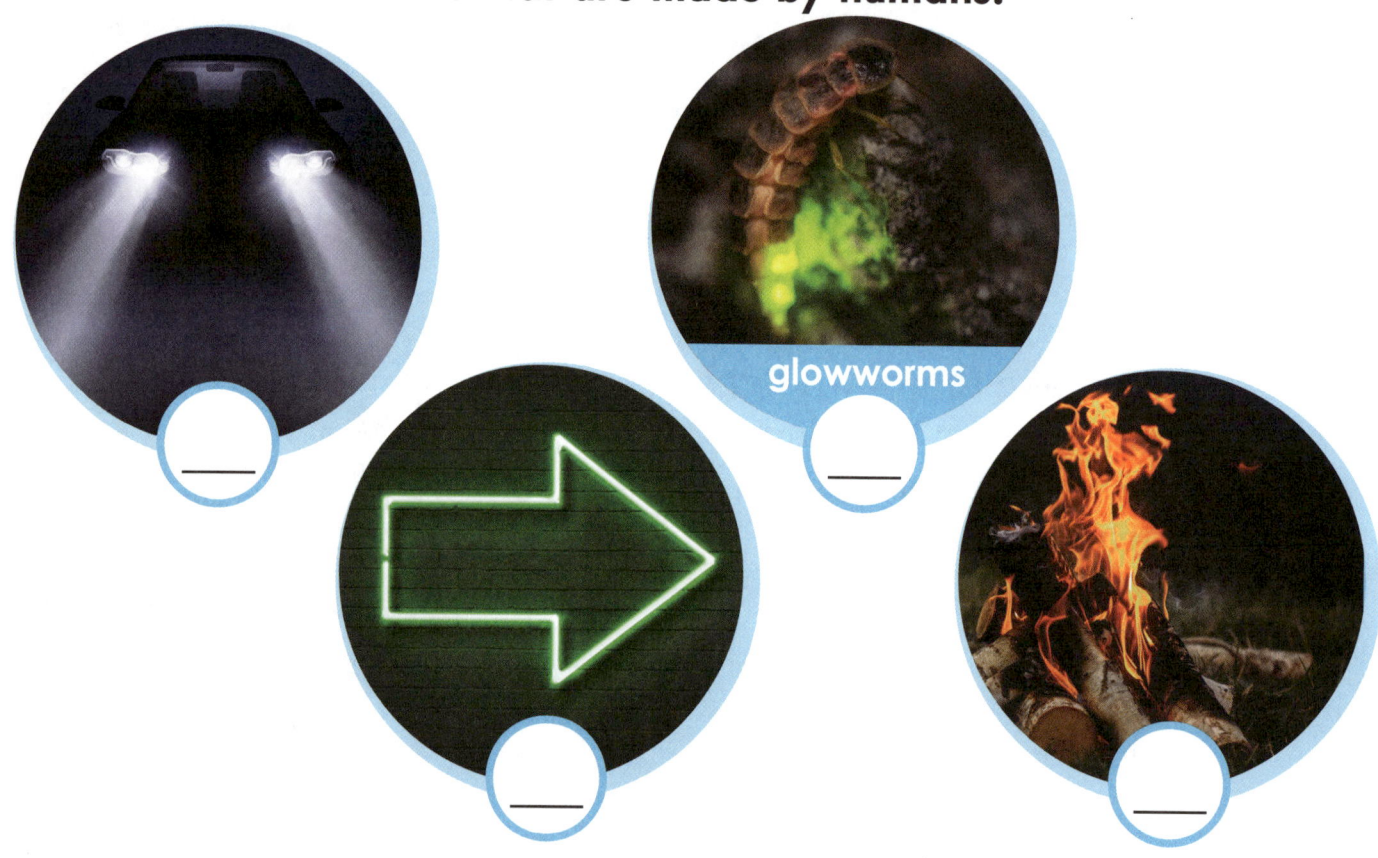

glowworms

____   ____   ____   ____

Fill in the circle next to the correct answer.

3. Choose a natural thing that allows you to see outside at night.
   ○ street lamp   ○ moon   ○ lit candle

4. Choose a thing made by people that helps you see outside at night.
   ○ flashlight   ○ fireflies   ○ stars

5. List two things you can see right now that do not make their own light.

_____   _____

6. Where is the light coming from that allows you to see these things? _____

116

Name _____

How Light Travels

What are shadows?

God made rays of light to move in straight lines. The rays keep moving in straight lines until they are blocked. A <mark>shadow</mark> is made when an object blocks light rays. The light makes an outline of the object.

1. Circle what makes light.

2. What blocks the light and makes a shadow? _____

The size and direction of a shadow depend on where the light is.

Your shadow looks small when the sun is right above you.

Your shadow looks tall when the sun is lower in the sky.

3. Draw the sun in the picture based on the ball's shadow.

4. Draw the house's shadow based on the sun's location.

Name _____

What Light Passes Through

Which items do not let light pass through?

God designed light to pass through items in different ways.

Some items let light pass through them.
Objects can be seen clearly on the other side.

Some items let a little bit of light pass through them.
Objects cannot be seen clearly on the other side.

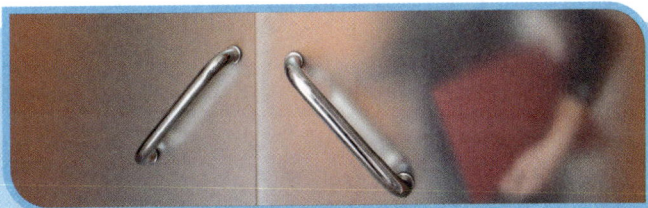

Some items do not let any light pass through them.
Objects cannot be seen at all on the other side.

1. Look at the three pictures. Which picture shows only a little light passing through an object? Circle it.

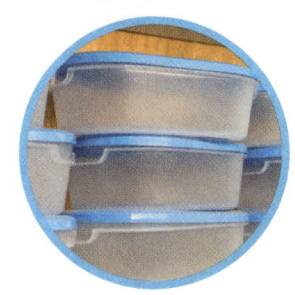  

Underline the word that best completes the sentence.

2. You cannot see objects clearly on the other side of the stained glass. The window lets ___ light through.

    all   some   no

3. You can see the fish clearly through the water. The water lets ___ light through.

    all   some   no

4. You cannot see what is behind the brick wall. The wall lets ___ light through.

    all   some   no

5. When the baby needs to nap during the day, the room needs to be dark like nighttime. Add something to the picture to help the baby sleep.

Name _____

Reflection

What reflects light?

God created light rays to reflect, or bounce back, when they shine on most objects.

Smooth, shiny objects reflect almost all light rays.

1. Circle the smooth objects that reflect light easily.

2. Make an **X** on the objects that do not reflect light easily.

Light colors reflect more light than dark colors.

3. Draw a square around the dog you could see better at night.

Wearing and using things that reflect light keep people safe.

Write Yes or No.

4. Dark colors reflect more light than light colors. _____

Write the answer on the line.

5. Would you be safer wearing light or dark colors outside at night?

_____

Name _____

Bending Light

What are some ways light can bend?

God created light to bend when it travels from one clear thing to another.

When light moves from air to water, the light bends. This bending can make things in the water look different.

Light bends through raindrops in the sky to create a rainbow.

Light can bend through other clear things to create rainbows.

Look at the penguin below. Its head does not seem to match its body.

Write the correct words on the lines.

1. The head is in the _____ and the body is in the _____.

2. The head and body look like they do not match because the light _____.

Lenses bend light and can make things look bigger.

3. Here is a ladybug. Draw how the ladybug looks when you look through the lens.

4. Make a check mark over the objects that use a lens or lenses.

5. Color the rainbow that this shape created.

Name _____

Communicating with Light

The power has gone out on your street. You need to communicate in the dark with your friend who lives next door. Your friend cannot hear or see you. Find a way to communicate well using light signals.

Materials
- flashlight
- cardboard
- red, blue, yellow cellophane

Step 1 State the Problem

I must communicate in the dark by using only light signals.

Step 2 Explore Ideas

Check the choices you might use.
☐ I can use number of flashes to show my answer choice of A, B, or C.
☐ I can use different colors to show my answer choice of A, B, or C.
☐ I can use different materials to show my answer choice of A, B, or C.

Step 3 Design a Plan

Complete the sentences.

1. I will use a flashlight and _____.

2. I will turn on the flashlight and place the _____ in front of the light.

3. I will use _____ for answer A,

   _____ for answer B, and

   _____ for answer C.

125

Step 4 Build and Test

1. Choose who will send the signal and who will watch the signal.
2. Have the sender listen to the teacher's question and signal an answer to the watcher.
3. Have the watcher tell the sender the answer that was signaled.

Step 5 Analyze and Redesign

Complete the exercises.

1. Did your partner get the signal correctly?   Yes   No

2. How can you make the signals better? _____
   _____
   _____

Make the changes and test again.

Step 6 Share Results

Share with the others whether you and your partner communicated well with your signal. State what kind of signal you used.

Complete the exercises.

1. What does God's design for light allow people to do?
   _____

2. Draw a way of using light to communicate.

Name _____

Chapter 7 Review

Write the correct word from the Answer Bank on the line.

ANSWER BANK   lines   shine   all   shadow   rainbow   see

1. Light allows people to _____.

2. God designed light to travel in straight _____.

3. If light is blocked, it creates a _____.

4. If something does not make its own light, light must _____ on it for it to be seen.

5. Items either let no light, some light, or _____ light pass through.

6. When light bends, it can create a _____.

7. Make an **X** on the sources of light made by God.

Write Yes or No on the line.

8. Your shadow looks big when the sun is right above you. _____

9. God created light to bend when it travels from one clear thing through another. _____

10. Dark colors reflect more light than light colors. _____

11. Smooth, shiny objects reflect almost all light. _____

12. Draw a square around the items that let all light through.

13. Match the word to its description.

lenses •   • the earth's main source of light

sun •   • make things look bigger because light bends

mirror •   • an object that reflects all light

Write the answer on the line.

14. Name an object that does not let all light pass through.

_____

15. Name something that makes light that is used to communicate.

_____

Unit Connect

Name _____

History
Kinds of Lights

God created the sun to give natural light.
People cannot use natural light all the time.
They had to find other ways to see in the dark.

1. Look at these lights. Which light would be easiest to use? Why?

_____
_____

torch   oil lamp   candle   lantern   light bulb

The light bulb is what most people use to see in the dark.
People keep making the light bulb better.
Older kinds of light bulbs get very hot. They do not last very long.
Newer light bulbs do not get as hot. They use less power and last longer.

oldest ➡ newest

2. Look around the room. What kinds of lights do you see?

_____

Technology
Hologram

A hologram is a type of picture made by casting light a special way.

Carmakers use holograms. Why would they use them?

Holograms help carmakers see all sides of the car at one time. Holograms help them see how to improve the car.

Doctors use holograms. Why would they use them?

Holograms help doctors see inside of a person. Holograms help them find what is wrong. Doctors look at holograms during surgery.

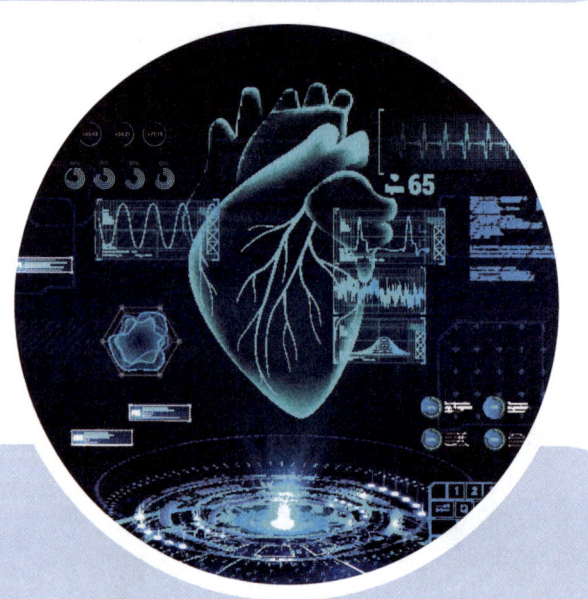

1. Who else might use holograms? _____

2. What would you make a hologram of? _____

Name _____

Window Shade

Sunlight shines through windows. The sun can shine in your eyes when you ride in a car. Plan and make a window shade to use in the car.

Step 1 State the Problem

Think about the problem you need to solve.

Sometimes the sun shines in my eyes when I ride in a car.

Step 2 Explore Ideas

Think about how a window shade works.

1. Choose how much light you want to pass through the shade.
   ○ I want to let some light pass.
   ○ I do not want any light to pass.

Think about what a window shade is made of.

2. How will you decide what to use?
   ○ I will test different materials.
   ○ I will use materials in my favorite color.
   ○ I will use the materials I like best.

3. Show how you will use a flashlight to test materials.

Step 3   Design a Plan

Design a window shade using materials that will let the right amount of light pass.

1. What light will you use to test items? _____
2. Test items and make an **X** to show how much light passes.

Light Test

Material	All light passes.	Some light passes.	No light passes.

Name _____

Window Shade

3. Draw your shade and label the items.

4. What is one item you will not use? Why?

_____

Step 4 Build and Test

Use your drawing to build your window shade.

1. Order the steps to build and test the shade.

___ Shine the flashlight at the shade.

___ Build the shade.

___ Look at how much light passes through the shade.

Use a flashlight to test your shade. Observe how much light passes.

A model is a small copy of something.
Models help you solve problems when the real thing cannot be used.
A flashlight is a model for the sun.

2. Compare a flashlight to the sun.

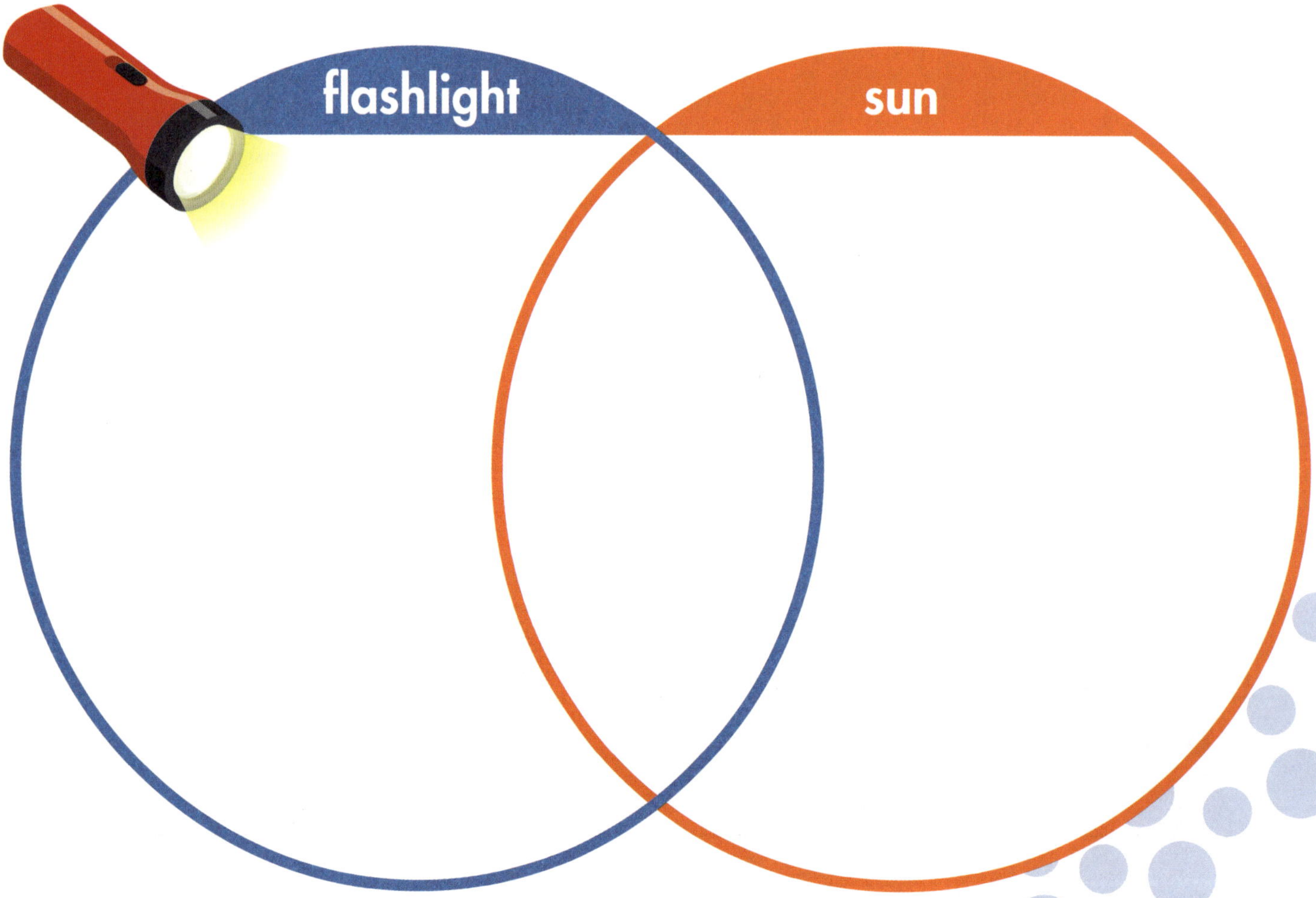

flashlight | sun

Name _____

Window Shade

Step 5 Analyze and Redesign

Analyze the results.

1. What is the shade supposed to do?
   ○ let all light pass
   ○ let some light pass
   ○ not let any light pass

2. What worked well?

   _____

   _____

3. What did not work well?

   _____

   _____

Analyze how the shade will work in a car. Redesign if needed.

Complete the exercises.

4. The car moves, and the sun comes in at different places. How will I hold the shade to always block the sun?

   _____

   _____

5. Where will the shade stay when I am not in the car?

   _____

   _____

Complete the exercises.

6. What else does the shade need?

_____

7. How will you change your design?

_____

8. Draw your new design and label the materials.

```
┌─────────────────────────────────────┐
│                                     │
│                                     │
│                                     │
│                                     │
│                                     │
│                                     │
│                                     │
└─────────────────────────────────────┘
```

Build and test your new design.
Complete the exercises.

9. Does it work?

_____

10. Do you need to make any other changes?

_____

11. What do you need that you do not have?

_____

Name _____

Window Shade

Step 6 Share Results

What can you tell others about your design?

1. Draw how to use the shade in a car.

Underline the words that describe your shade.

2. The shade lets **a little light** / **no light** pass.

3. It is **the same size as** / **smaller than** the window.

4. The shade **is** / **is not** a model.

5. Draw what the shade looks like when it is not in use.

Share your design solution with another group.
Complete the exercises.

6. What do you like about your design?

_____

7. What do you like about their design?

_____

Unit 4
Earth Science

He made the moon to mark the seasons,
and the sun knows when to go down.
Psalm 104:19

Name _____

Objects in the Sky

Chapter 8

Where on Earth are these from?

Science Grade 1

141

They are not from Earth at all. These are craters, or big dents, on the moon. Giant rocks, called asteroids and meteors, hit the moon and make craters. Bigger rocks make bigger craters. Smaller rocks make smaller craters.

Make a crater for each asteroid.

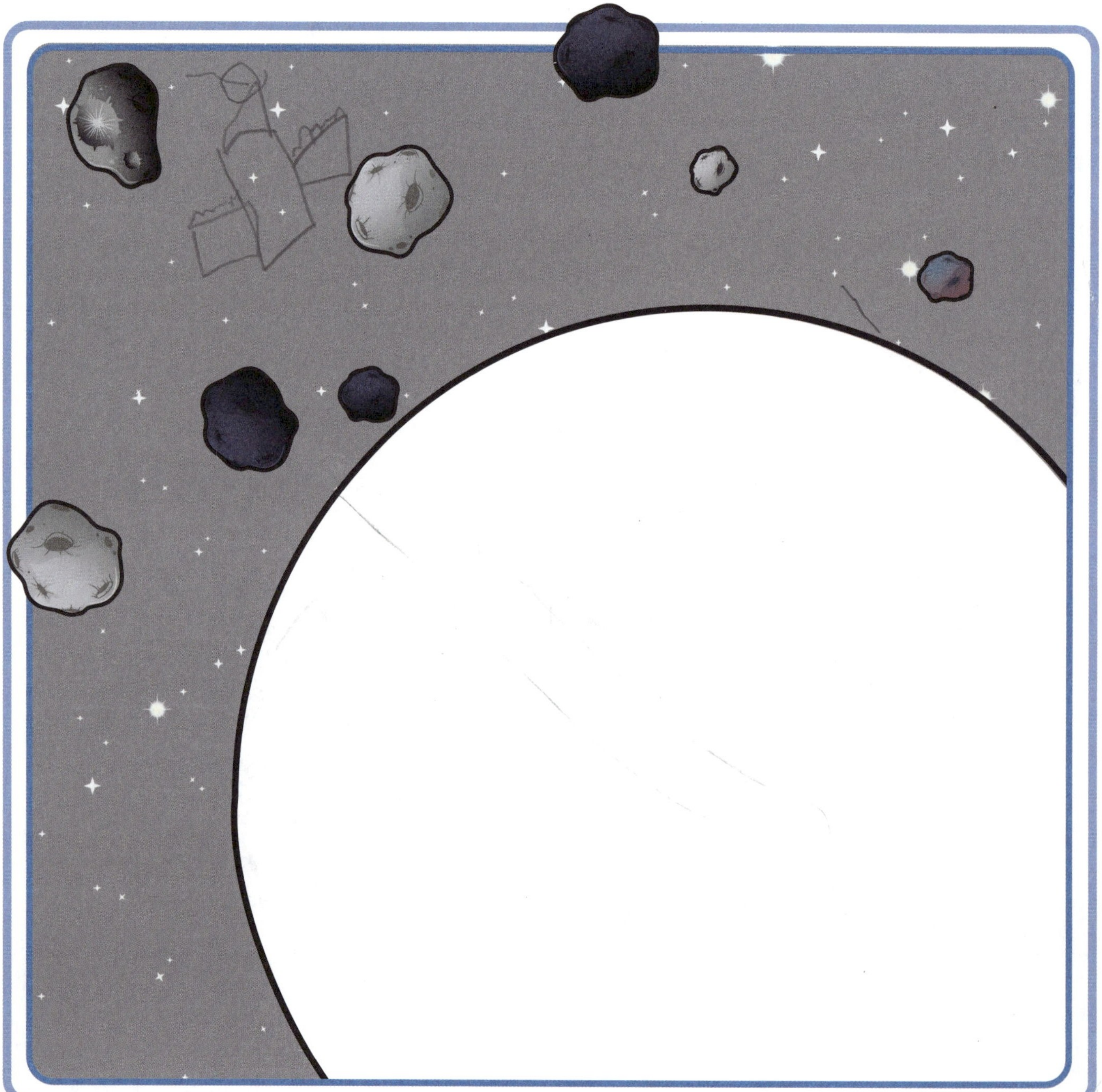

142

Name _____

Day and Night

What makes day and night?

Earth spins in a full turn each day. It follows the same pattern every day. This pattern makes day and night. Because Earth spins, the sun seems to move across the sky.

The drawing taped to the balloon is a model of a person on Earth.

1. What is the flashlight a model of?

   _____

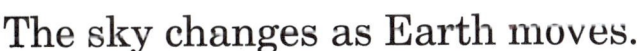

The sky changes as Earth moves.

Observe whether the sun looks high or low in the sky. Circle the answer.

2. high   low   3. high   low   4. high   low

morning   noon   evening

145

Light comes from the sun. The house blocks the sunlight from hitting the ground. The house makes a shadow. As the sun seems to move across the sky, the shadow moves too.

5. Draw arrows to show the way the sun seems to move during the day.

Look at the pictures above and underline the correct answer.

6. When is the shadow the shortest?
morning   noon   evening

The children block the sunlight from hitting the ground. They make shadows.

Look at the shadows to complete the exercises.

7. Check the box for the picture that is closest to noon.

8. How do you know? _____

Name _____

8.3

Moon and Stars

Why do you see the moon and stars at night?

God made the moon and the stars to give some light at night.

The moon cannot make its own light. The moon reflects the light of the sun.

The moon looks different each night. Some nights, only part of the moon is showing.

The night sky is full of stars. There are too many to count. Stars make their own light. But they do not look bright because they are very far away.

Use the pictures above to complete the exercises.

1. Draw one star beside the picture of the whole moon showing.

2. Draw two stars beside the picture of part of the moon showing.

The moon and stars are far away. Scientists use a **telescope** to study them. A telescope makes objects in the sky look closer and larger.

3. Circle the picture that shows the moon through a telescope.

4. Use words from the Answer Bank to compare the moon and stars.

| ANSWER BANK | makes its own | reflects | many | one | different | the same |

	moon	stars
how many there are	_____	_____
how it shows light	_____	_____
how it looks each night	_____	_____

148

Name _____

8.4

Patterns in the Night Sky

What kinds of patterns are in the night sky?

The moon is round like a ball. Its shape does not change. But it looks different each night. Some nights, the whole moon can be seen. Other nights, the moon does not show at all. The look of the moon follows a pattern each month. Observe the pattern in May and June.

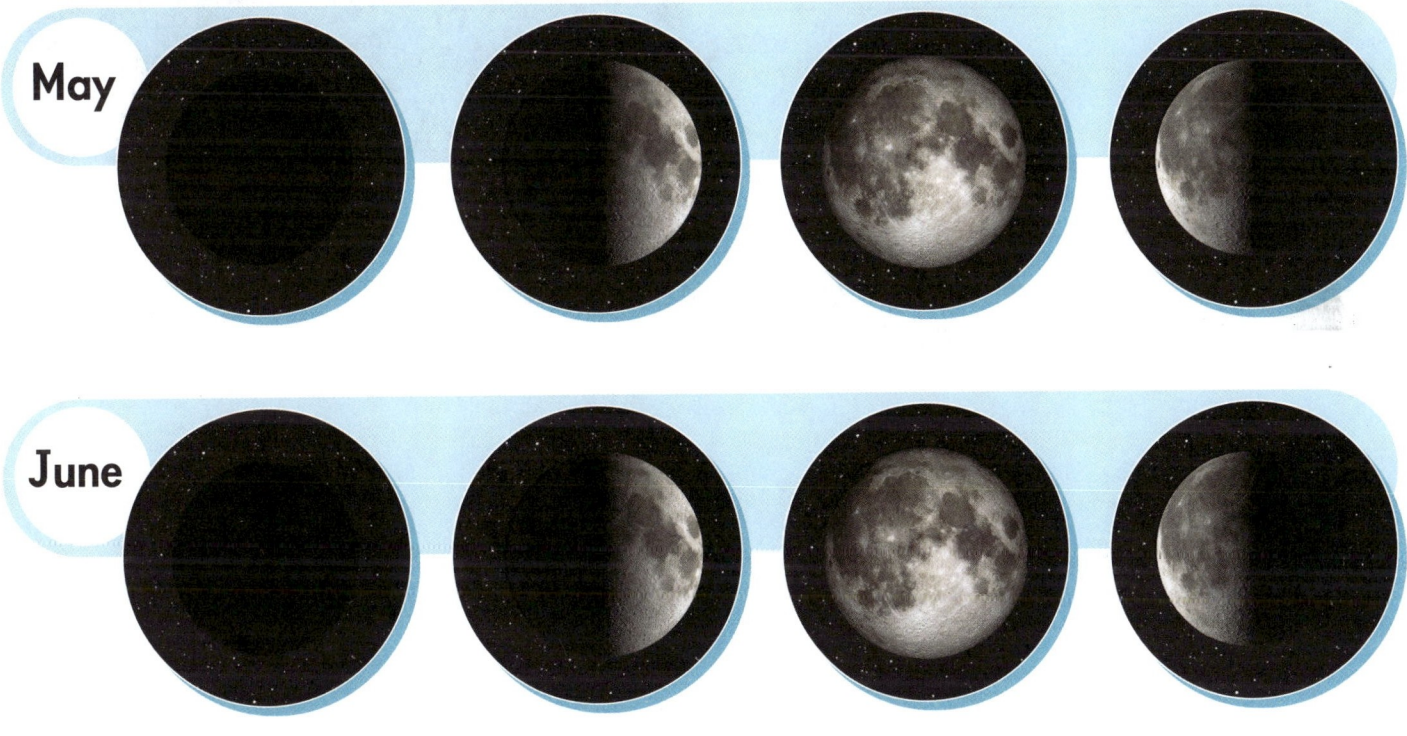

1. Color to show how the moon will look in July.

2. How do you know what the moon will look like in July?

_____

149

You can see many stars in the sky at night. Have you ever tried to count the stars?

Groups of stars form patterns.

3. Why do you think this group of stars is called the Big Dipper?

_____

_____

_____

This group of stars makes a bear.

4. Connect the stars to make the shape of the Bear. Do not cross any lines.

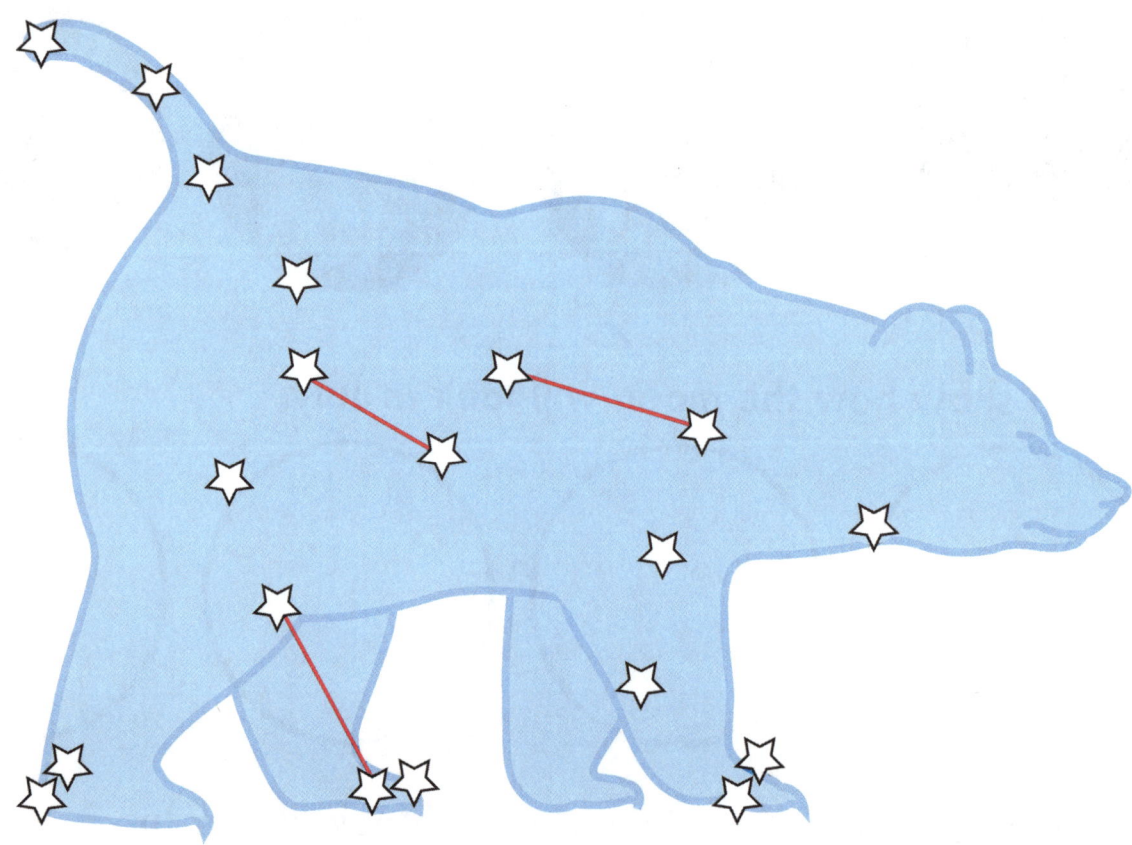

5. Who made the moon and stars to have order? _____

Name _____

8.5

Differences in the Day and Night Skies

What can you find in the day and the night skies?

The sun is seen only in the day sky.

The stars are seen only in the night sky.

Some things can be seen in the day sky and the night sky.

1. Write the words in the correct place on the diagram.

ANSWER BANK: sun   moon   stars   clouds   plane   songbird   bat

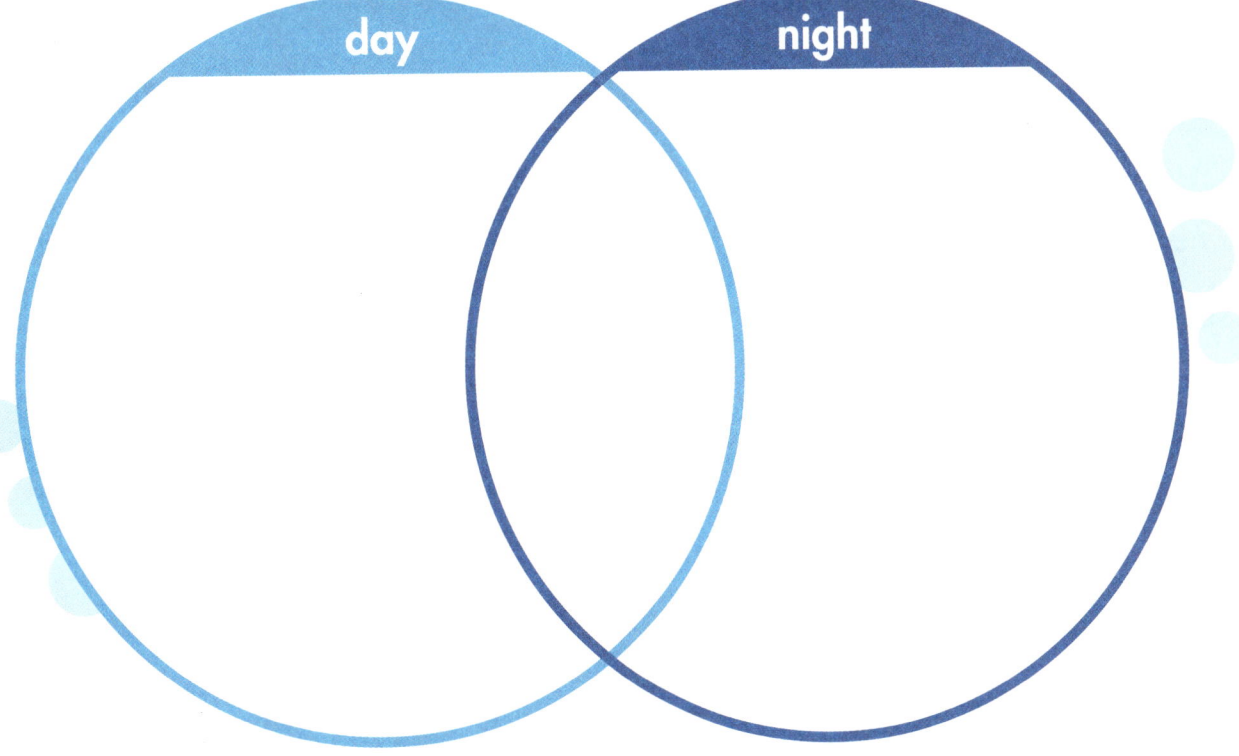

day   night

151

The sun seems to move from east to west every day. It heats Earth in the day time.

The moon moves and changes the same way each month. It reflects the light of the sun. It does not heat Earth.

Fill in the circle for the correct answer.

2. The sun and the moon both ___
○ are in the night sky.
○ make their own light.
○ have patterns in the sky.

3. God is creative. He made different ___ for day and night.
○ lights   ○ moons   ○ stars

4. Make a check mark under what is true for each object.

	seen at daytime	seen at nighttime	makes light	shows different shapes
sun				
moon				
star				
cloud				

Name _____

Investigation 8.6

Shadows

Step 1 — Ask a Question
Where will today's afternoon shadow be compared to the afternoon shadows on other days?

Step 2 — Make a Prediction
Make a check mark for your prediction.
The shadow will be in __ other days.

☐ the same place as   ☐ a different place from

Materials
- pictures of an object's shadow taken during the morning and afternoon over 3 days

Step 3 — Plan and Do a Fair Test
Follow the steps to test your prediction.

1. Go outside at two set times each day for 3 days.
2. Record where the shadow of the same object falls at each time.
3. Compare the Day 3 shadows with the Day 1 and Day 2 shadows.

Step 4 — Record and Analyze Results

1. Use a black crayon. Mark the shadows for Day 1 and Day 2.
2. Use a red crayon. Mark the shadows for Day 3.

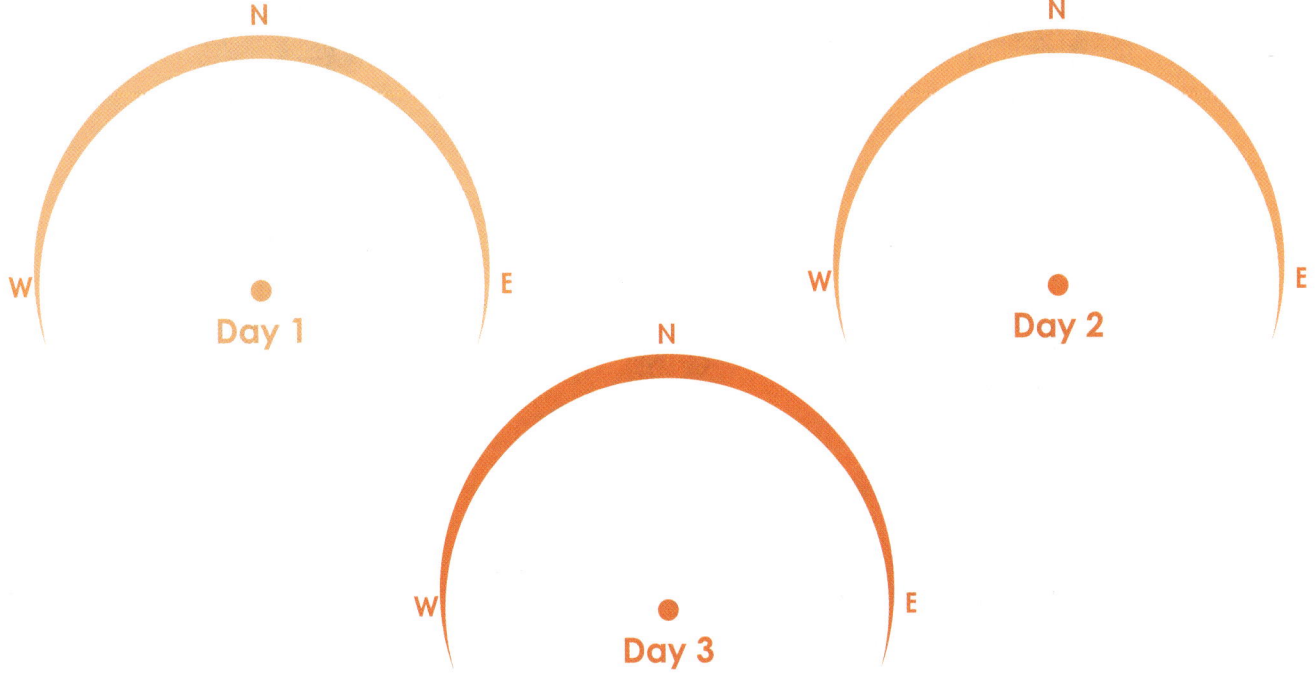

Day 1   Day 2   Day 3

153

Step 5 Make a Claim

Circle your answers.

1. Where is the Day 3 afternoon shadow compared to the other afternoon shadows?

in a different place   in the same place

2. Where will the morning and afternoon shadows be tomorrow?

in different places   in the same places

Step 6 Give Evidence

How can you know where a shadow will be at a given time and place?
Make a check mark.
I can know because the sun moves
☐ the same way every day.
☐ in a different way every day.
☐ more every day.

Step 7 Share Results

Ben waits for the bus at the same time and place every day.
Complete the exercises.

1. Draw Ben's shadow on the third day.

2. I know where to draw the shadow because the **sun / bus** moves the same way every day.

154

Name _____

Chapter 8 Review

1. Look at the pictures of sunlight coming in the windows. Make a check mark on the picture that was taken at noon.

Color the correct objects.

2. God made ___ to light the day.

3. God made ___ and ___ to light the night.

4. Which object does not make its own light?

Look at the picture of Earth.

5. Label the sides **day** and **night**.

6. How do you know which side is day and which side is night?

_____

_____

155

Fill in the circle of the correct answer.

7. Which tool do scientists use to study the night sky?
   ○ telescope   ○ hand lens   ○ sunlight

8. Which looks like it changes shape each night?
   ○ sun   ○ moon   ○ stars

9. Which makes its own light in the night sky?
   ○ sun   ○ moon   ○ stars

10. Which does not have a pattern in the night sky?
   ○ sun   ○ moon   ○ stars

Complete this night picture.

11. Add something that can only be seen in the night sky.
12. Add something that can be seen in the day sky and the night sky.
13. Color the sky.

Name

Weather and Seasons

Chapter 9

How are snowflakes made?

Water in a cloud can freeze and turn to ice. The little pieces of ice join together with other pieces of ice and form a snowflake. Ice can come together in many ways making special snowflake designs. A snowflake falls when it gets too big and heavy.

Draw your own snowflake.

Name _____

9.1

Weather

What are different kinds of weather?

Weather is what the air is like outside. God made all kinds of weather. Some days may be sunny. Some days may be cloudy. Other days may be windy, rainy, or snowy.

1. Use the Answer Bank to write the kind of weather under each picture.

ANSWER BANK: snowy   rainy   cloudy   windy   sunny

159

Some weather can create storms. If you see dark clouds, a storm may be coming. Storms are not safe.

A thunderstorm has rain, strong wind, thunder, and lightning.

A tornado is a strong, twisting windstorm.

A hurricane forms over ocean water. It has lots of rain and strong winds.

A blizzard is a storm with **strong winds and snow.**

Circle the correct answer.

2. A blizzard has strong winds and lots of ___.

rain   snow   sun

3. A hurricane has strong winds and lots of ___.

rain   snow   sun

4. A strong, twisting windstorm is called a ___.

lightning   thunderstorm   tornado

5. ___ are a sign that a storm may be coming.

Sunny skies   Dark clouds   Gentle winds

Name _____

Measuring Weather

How is weather measured?

Many kinds of tools measure weather. A <mark>thermometer</mark> measures how hot or cold something is.

You can place a thermometer outside to see how hot or cold it is. If the thermometer reads 90°F, it is a hot day. But 20°F is a cold day and may even be snowy!

Read the number where the red line stops.

1. What is the temperature? _____ °F

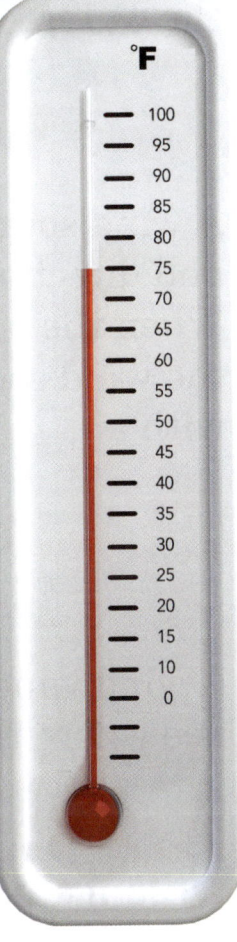

Color the thermometer to show the temperature.

2. 40°F

3. 60°F

There are many other tools that measure weather.

A rain gauge measures how much rain has fallen. To measure the rain, look at the line closest to the water.

A wind sock shows the speed and direction of the wind. When the wind blows fast, the wind sock lifts and fills with air.

A wind vane tells the direction of the wind. See where the arrow is pointing.

Write the answer on the line.

4. How much rain fell? ___ inches

5. Is the wind blowing fast or slowly? _____

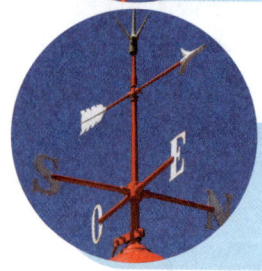
6. What is the direction of the wind? _____

Name _____

9.3

Seasons

What is different in each season?

A <mark>season</mark> is a time of year. God created the seasons. There are four seasons in some parts of the world. Other parts of the world have only a wet season and a dry season. All seasons occur in a cycle or a pattern. Some seasons are colder or warmer than others.

Fill in the circle for the correct answer.

1. ___ is the coldest of the seasons.
   ○ Spring   ○ Summer   ○ Autumn   ○ Winter

2. ___ is the hottest season.
   ○ Spring   ○ Summer   ○ Autumn   ○ Winter

Fill in the circles for all the correct answers.

3. ___ is warmer than winter.
   ○ Spring   ○ Summer   ○ Autumn   ○ Winter

4. ___ is cooler than summer.
   ○ Spring   ○ Summer   ○ Autumn   ○ Winter

Sunlight hours also follow a pattern. Each season has different hours of sunlight. The cycle repeats each year. This graph shows about how much sunlight a day of each season has in Washington, DC.

Use the graph to answer the questions.

5. Which season has more sunlight, spring or summer? _____

6. Which season has the least? _____

7. If you wanted to play outside for the longest, which season would you choose? _____

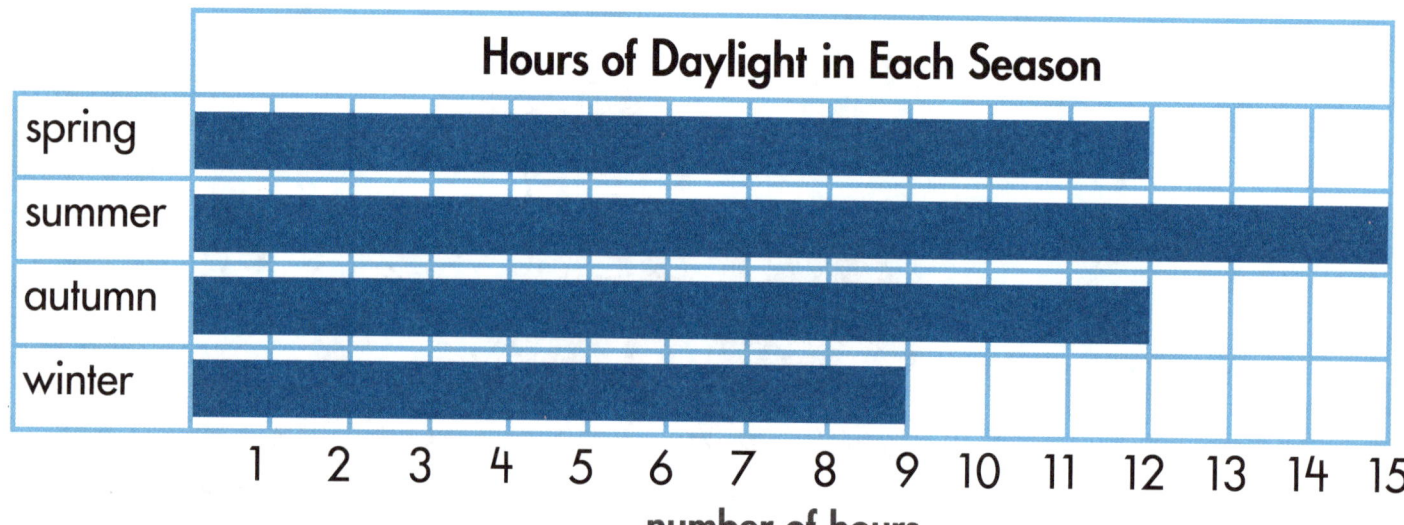

In cool seasons, people wear thick, warmer clothes to keep warm. In warm seasons, people wear thin, cooler clothes to stay cool.

8. Draw clothes on the boy that will keep him warm.

Name _____

9.4

Seasons and Plants

What happens to plants during the seasons?

God made plants to cycle through the seasons.

The oak tree loses leaves in autumn. It has no leaves in the winter. It grows new leaves in the spring. By the summer, it is full of leaves. It repeats this cycle year after year.

1. Write the correct season on the line.

_____

_____

_____

_____

Most plants go through their life cycles within a year.

1 Flowers are planted in the spring and start to grow.

2 They keep growing all summer long.

3 In autumn, when the first frost comes, some plants start to die.

4 In the winter their roots, stems, and leaves die.

Some plants do not fully die in the winter. Only their top part dies. In spring, the plant regrows from the same roots.

Complete the exercises.

2. Why do some plants die in autumn?

_____

3. Why would a plant not grow well if it was planted in the winter?

_____

4. During which season are most new plants the smallest? _____

Name _____

9.5

Seasons and Animals

What do animals do in each season?

God gave animals different ways to live in each season.

Some animals are less active in the winter. They sleep more.
Some bats sleep in winter.
Other bats fly from place to place to find food.

Some animals' fur changes color to blend in with their environment.
Other animals get thicker fur in the winter.

1. Make a check mark by the animal that blends in with winter snow.

In spring, the weather gets warm. Food starts to grow.
Many animals have babies.

2. Draw a baby animal in the box.

3. What baby animal did you draw?

_____

167

In the summer, animals are very active. Some lose their thick winter fur to keep cool.

In autumn, some animals like geese move to warmer places. Others gather and store food for the winter.

4. Circle the animal that is gathering food.

5. Make a check mark in the square that tells how each animal survives the winter.

	sleeps more	moves to a warmer place	stores food	changes color of fur

Name _____

How the Sun Warms Surfaces

Investigation 9.6

Step 1 Ask a Question

Does sunlight _____ ice fastest on

_____, _____, or _____?

Materials
- plate of sand
- plate of soil
- plate of rocks
- 3 ice cubes
- ruler

Step 2 Make a Prediction

Fill in the circle above your choice.
I think sunlight melts ice fastest on __.

○      ○      ○

Step 3 Plan and Do a Fair Test

Do the test. Write measurements in the Step 4 chart. Check off each part after you do it.

1. ☐ Measure and record the height of an ice cube.

2. ☐ Place the ice cube on the plate of sand.

3. ☐ Measure and record the height of another ice cube.

4. ☐ Place the ice cube on the plate of soil.

5. ☐ Measure and record the height of the last ice cube.

6. ☐ Place the ice cube on the plate of rocks.

7. ☐ Set the 3 plates with ice cubes directly in the sun.

8. ☐ Every 10 minutes for 30 minutes measure the height of each ice cube.

Step 4 — Record and Analyze Results

Write the measurements in the appropriate box.

	0 minutes	10 minutes	20 minutes	30 minutes
sand	_____ cm	_____ cm	_____ cm	_____ cm
soil	_____ cm	_____ cm	_____ cm	_____ cm
rocks	_____ cm	_____ cm	_____ cm	_____ cm

Step 5 — Make a Claim

The sunlight melted the ice fastest on the _____.

Step 6 — Give Evidence

How do you know your claim is true?

_____

_____

Step 7 — Share Results

Share your claim and evidence with another group. Answer the questions.

1. How are your results the same as the other group's results?

   _____

2. How are your results different from the other group's results?

   _____

3. What did God design the sun to do to Earth's surface?

   _____

Name _____

9.7

Chapter 9 Review

Write the correct words from the Answer Bank on the lines.

ANSWER BANK: thermometer   weather   sun   blizzard

1. A forecast tells what the _____ will be.

2. A _____ measures how hot or cold something is.

3. A _____ is a storm with lots of snow and strong wind.

4. The _____ warms Earth.

5. Draw a square around the clothes that are best for a rainy season.

6. Draw a line to match the type of weather to its description.

 windy • • makes Earth's surface warm

 snowy • • creates a lot of shade

 sunny • • makes trees sway

 cloudy • • cold and happens mostly during winter

7. Draw what a tree looks like during autumn.

Write Yes or No.

8. Evergreen trees lose all their leaves during autumn. _____

9. Some places have a wet season and a dry season. _____

10. Many animals have babies in the autumn. _____

Answer the questions.

11. Which season has the longest daylight hours? _____

12. In which season do most flowers begin to grow? _____

13. Make an **X** on the animals that are not getting ready for winter.

storing food

flying to a warmer place

playing

Name _____

Unit Connect

Career

Meteorologist

Meteorologists study the weather. They use technology to predict the weather. Their predictions keep people safe.

Doppler radar tower

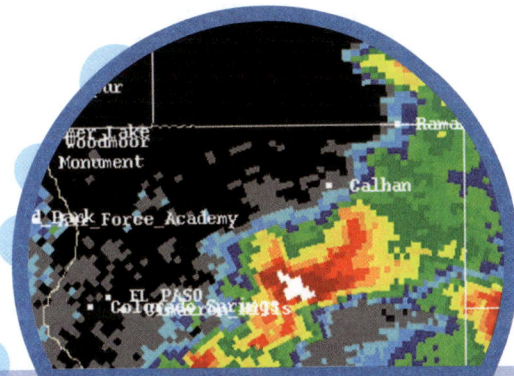
Doppler radar image

weather satellite

satellite image of a hurricane

1. How do meteorologists help people? _____

_____

Unit Connect

Biography
Robert Fitzroy

Robert Fitzroy was the first meteorologist to make weather forecasts. He printed the forecasts in a newspaper.

Later, other meteorologists printed forecasts in newspapers too. Now people have forecasts on televisions and cell phones.

Use the forecast to answer the questions.

1. How many days of the week are not sunny? _____

2. What kind of weather will Thursday have? _____

3. Which days of the week should you wear a raincoat?

_____

Name _____

Wind Sock

You want to fly your kite. But you need to know where the wind is blowing from and if it is blowing fast enough. Create a wind sock that shows the wind's direction and how fast it is blowing.

Step 1 State the Problem

Think about the problem you need to solve.

I need to know the direction and speed of the _____

so I can fly my _____.

Step 2 Explore Ideas

Think about the shape of a wind sock.

A wind sock is a cone shape. Both ends are in the shape of a circle. One end is bigger than the other.

1. Trace the wind sock.

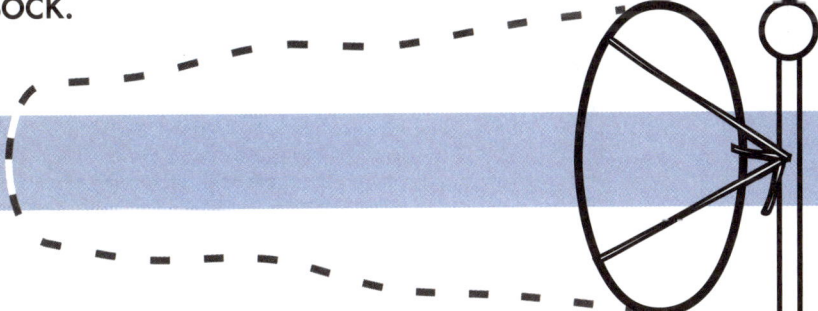

Think about how a wind sock works.

Wind travels into the larger opening and through the tube. The smaller end of the tube points in the opposite direction the wind is coming from.

2. The wind sock is pointing east. What direction is the wind coming from?

_____

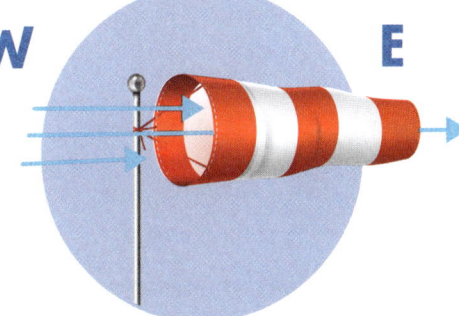

Think about what a wind sock is made of.
Wind socks are made of light material.

3. Circle the lightest material.

Think about the color of a wind sock.
Wind socks are bright colors so people can see them.

4. Make a square around the color that is bright and easy to see.

Step 3 Design a Plan

Design a wind sock using materials that are bright enough to see. Pick materials that are light enough for the wind to easily move.

1. List the materials you will use. _____

_____

_____

2. Draw a picture of your design. Label the materials.

Name _____

Wind Sock

Step 4 Build and Test

Build your wind sock to look like your drawing.

1. Draw the wind sock after you have built it.

2. Was your design easy to make? _____

3. What was the hardest part to make? _____
_____

Test your wind sock. Place it outside and observe what it does.

4. Draw how the wind sock looks in the wind.

Step 5 Analyze and Redesign

Analyze the results.

1. Did your design work? _____

2. How do you know? _____
   _____

3. Which direction is the wind sock pointing? _____

4. Which direction is the wind coming from? _____

5. Color between the lines to show how far the wind sock sticks out.

6. Circle how fast the wind is blowing.

   slowly   a little fast   very fast

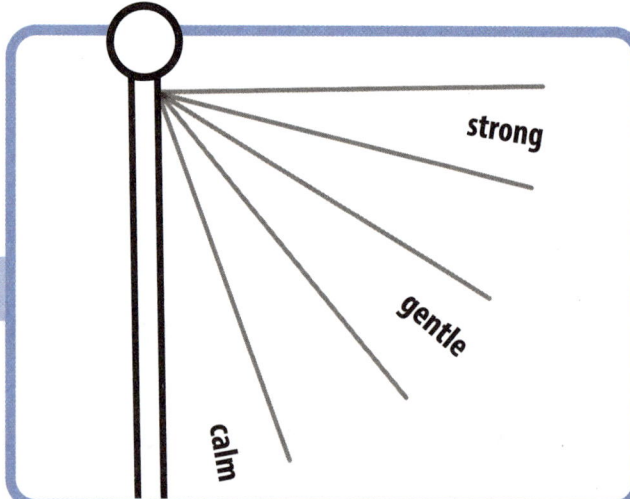

Redesign.

7. What would you do to change your design? _____
   _____

Step 6 Share Results

Compare your design and results with another group.

1. What materials did they use that you did not? _____
   _____

2. Did their wind sock show the same speed and direction as yours?
   _____

Unit 5
Human Body

Let the peace of Christ rule in your hearts, since as members of one body you were called to peace.
Colossians 3:15a

Name _____

Five Senses

Chapter 10

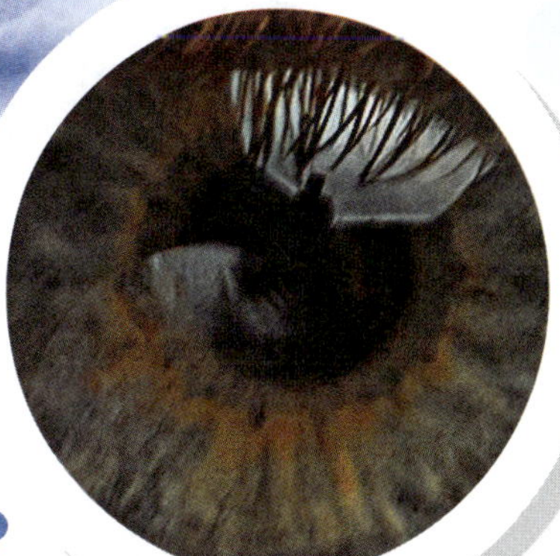

What is this picture?

© Science Grade 1

This picture is of a human eye.

Human eyes can be different shades of blue, green, or brown.

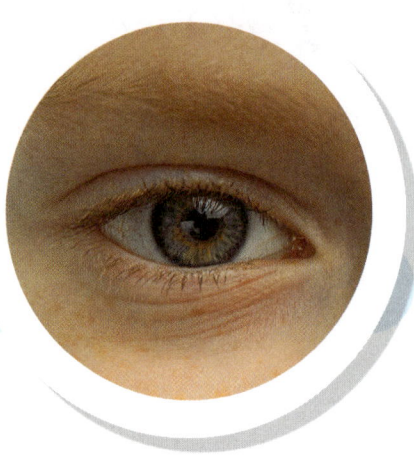

Animal eyes can be a different shape and color than human eyes. Animal eyes can even be black, red, yellow, orange, or white.

Color the eye to look like your eyes.

Name _____

Sight and Hearing

How do people see and hear?

God gave people senses that help them enjoy the world. Senses help people learn about the world. The senses are sight, hearing, touch, smell, and taste.

The human body is the whole physical self. People use parts of the body for each sense. Eyes let people see. They see color, shape, size, and texture.

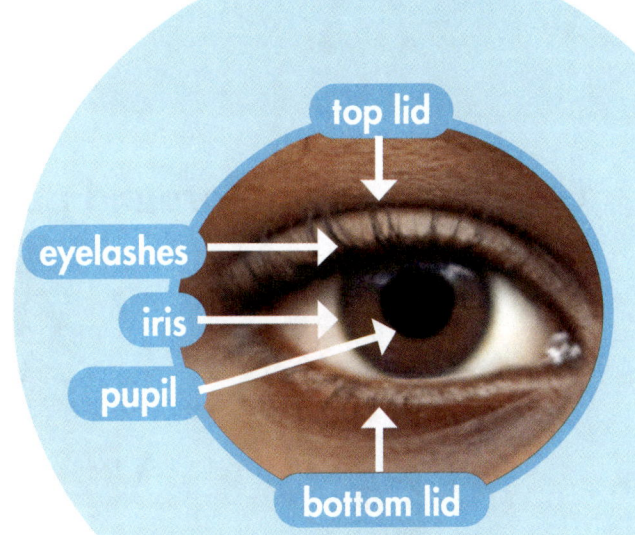

Use your eyes to see the picture of a room.

1. Make an **X** on something that is green.
2. Draw a circle around something that is square.
3. Make a check mark on something that is soft.
4. Draw a square around something that is smooth.

Fill in the circles next to all the correct answers.

5. Eyes let you see ___.

    ○ colors   ○ sounds   ○ shapes   ○ textures

183

Ears let people hear sounds. Sounds can be loud or soft. They can be high or low.

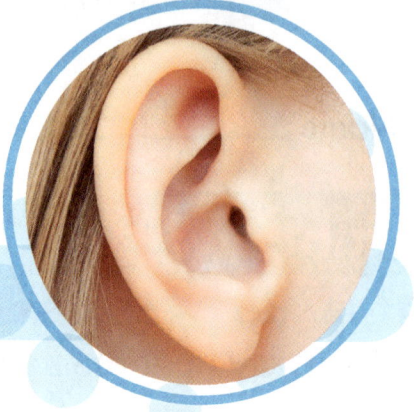

6. Draw a square around people or things that are making a sound.

7. Draw something from your home that you like to hear.

Name _____

Touch

How do people feel things?

Skin is the outer cover of people's bodies. Skin touches and feels things.

Hands, feet, and lips can feel things.

1. Write a sentence about something you touch with your hands, feet, or lips. _____

_____

Write **hands**, **lips**, or **feet** to complete the sentence.

2. You can use your _____ to feel a dog when you pet it.

3. You can use your _____ to feel the sand when you walk on the beach.

4. You can use your _____ to feel a warm drink.

People can feel different shapes and sizes.

5. Which object feels large, smooth, and round? Circle it.

People can feel different textures.

6. Draw a line from each item to how it feels.

7. Write words from the Answer Bank that tell about each animal.

ANSWER BANK
- fluffy
- smooth
- furry
- slimy
- warm
- cold

8. Write a sentence about one animal on this page. Use words from the Answer Bank.

_____

Smell and Taste

How many types of smells are there?

God gave people a sense of smell. They can enjoy many kinds of smells.

Flowers have a fragrant smell. Lemons and oranges have a citrus smell. Other fruits have fruity smells. Chocolate and vanilla smell sweet. Peppermint candy smells minty. Fresh cut grass is a kind of woody smell.

1. Draw and color the item next to its smell.

   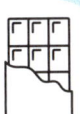

fragrant	
citrus	
minty	
sweet	

2. What is your favorite smell?

_____

3. Write a smell you do not like.

_____

God gave people a sense of taste. People use their tongues to taste things.

Foods can taste sweet if they have sugar in them. Ice cream is sweet.
Foods may taste salty if they have salt on them. Chips are salty.
Foods like lemons may taste sour.
Kale and coffee taste bitter.

4. Label each food as sweet, salty, sour, or bitter.

cranberries

limes

_____ _____

cupcake

pretzels

_____ _____

5. What is your favorite taste? _____

Name _____

Human and Animal Senses

How are human senses different from animal senses?

God gave animals and humans senses. Some animals have stronger senses than people.

Hawks can see small things from farther away than people can.

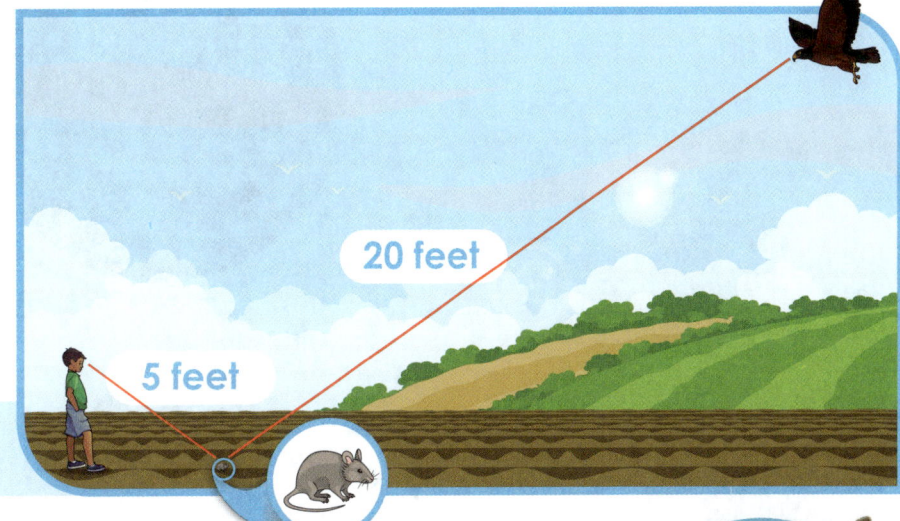

Cats can see clearly in the dark, but humans cannot. Cats can also hear high sounds that people cannot hear.

Sometimes animals use different body parts than humans to sense things. A platypus uses its bill for touch. Its bill can tell how hot or cold something is.

Underline the correct word to complete the sentence.

1. A **human / hawk** can easily see small things from 20 feet away.

2. A **human / cat** can see clearly at night.

3. A **human / platypus** uses hands for touch.

Most animals use their noses to smell. Some animals can smell things from farther away than people can.

Most animals use their tongues to taste. Some animals have more taste buds than humans. Catfish have ten times more taste buds than people.

4. Read about the tiger. Complete the chart about humans and tigers by circling the correct word.

A tiger uses it eyes to see. It can see clearly at night.
A tiger's whiskers are for touching.
A tiger has two ears that turn toward sounds.
A tiger uses its nose to smell.
A tiger tastes things that are salty, bitter, and sweet with its tongue.

human	A human **can / cannot** see clearly at night. Humans use **hands / whiskers** to touch.
tiger	A tiger **can / cannot** see clearly at night. A tiger uses its whiskers to **touch / smell**.
both	Both use their tongue to **hear / taste**. Both have two **ears / tails** and two **feet / eyes**.

Name _____

Senses and Safety

What do senses do for humans?

God gave people senses for their safety. Senses help keep people free from getting hurt. On a busy street, sight and hearing keep people safe. Look for and listen to what is around you, like cars, signs, and people.

Look at signs to know when it is safe to cross a street.

Listen to crossing guards.

Circle what to see and hear to stay safe.

1.

2.

The senses of smell and taste can also keep people safe. Food may smell and taste bad when it is no longer safe to eat.

Some gases smell bad and warn of danger. What might a smoky smell warn you about?

The sense of touch keeps people safe. If it is dark, a person could feel for a light switch or a flashlight. People's sense of touch can tell the difference between objects without seeing them.

3. Which senses can help you find your dog in the dark? Circle them.

hearing   taste   sight   touch

4. Color each item to match the sense that tells you of danger.

Name _____

Food and Senses

Investigation 10.6

God gave people the five senses to enjoy His creation. What about food makes eating it enjoyable?

Materials
- blindfold
- paper plate
- piece of apple or piece of pear

Step 1 Ask a Question

Do sight and smell affect taste?

Step 2 Make a Prediction

Circle your prediction.

I think sight and smell will affect taste. **Yes / No**

Step 3 Plan and Do a Fair Test

Use the picture to order the steps.

___ Tell what food was tasted.

___ Pinch your nose shut.

___ Taste a piece of food from your teacher.

___ Put on a blindfold.

Step 4 Record and Analyze Results

Finish the sentences.

1. The food tastes like _____.

2. The food is really _____.

3. Tally the results from the whole class.

Results	Tallies	Number
identified the apple correctly		
identified the apple incorrectly		
identified the pear correctly		
identified the pear incorrectly		

4. Color the bar graph to match the tally marks.

Class Taste Test Bar Graph

identified the apple correctly													
identified the apple incorrectly													
identified the pear correctly													
identified the pear incorrectly													

0 1 2 3 4 5 6 7 8 9 10 11 12
number of students

Step 5 — Make a Claim

Circle your claim.

Sight and smell **do / do not** affect taste.

Step 6 — Give Evidence

How do you know that your claim is true? _____

_____

Step 7 — Share Results

What different foods could you use to try this activity at home?

Name _____

10.7

Chapter 10 Review

Use the Answer Bank to complete the sentences.

ANSWER BANK: pupil   safety   sight   smell   taste   hearing

1. The sense of _____ lets people know if food is salty.

2. The sense of _____ lets people sniff different scents.

3. The sense of _____ lets people listen to a bird sing.

4. The sense of _____ lets people see colors.

5. The part of the eye that lets in light is the _____.

6. God designed the senses for people to enjoy things and for _____.

7. Label each body part and the sense that goes with it.

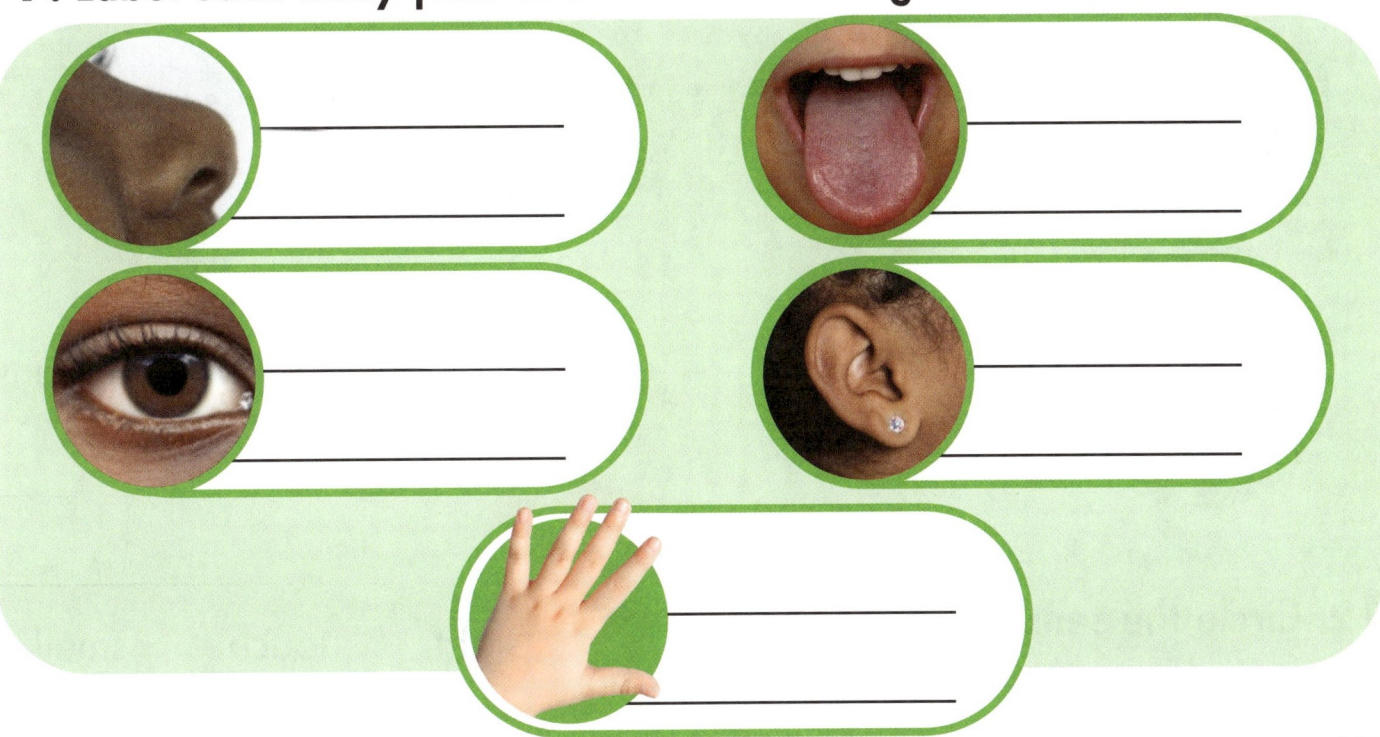

Pretend the cookie is real. Answer the questions.

8. What colors do you see?

_____

9. What texture does the cookie have? _____

10. What kind of taste does it have? _____

11. Draw a line from the animal to the description of its sense.

   •   •   can hear a high pitched noise that humans cannot

   •   •   uses its nose for touch

   •   •   can see farther away than humans

12. Circle the senses that work with taste to make eating more enjoyable.   sight   touch   smell

196

Name _____

Heart and Lungs

Chapter 12

What can you do with your breath?

One thing you can do with your breath is blow out candles.

Draw the number of candles that you will have on your next birthday cake. Draw a picture of yourself blowing out the candles.

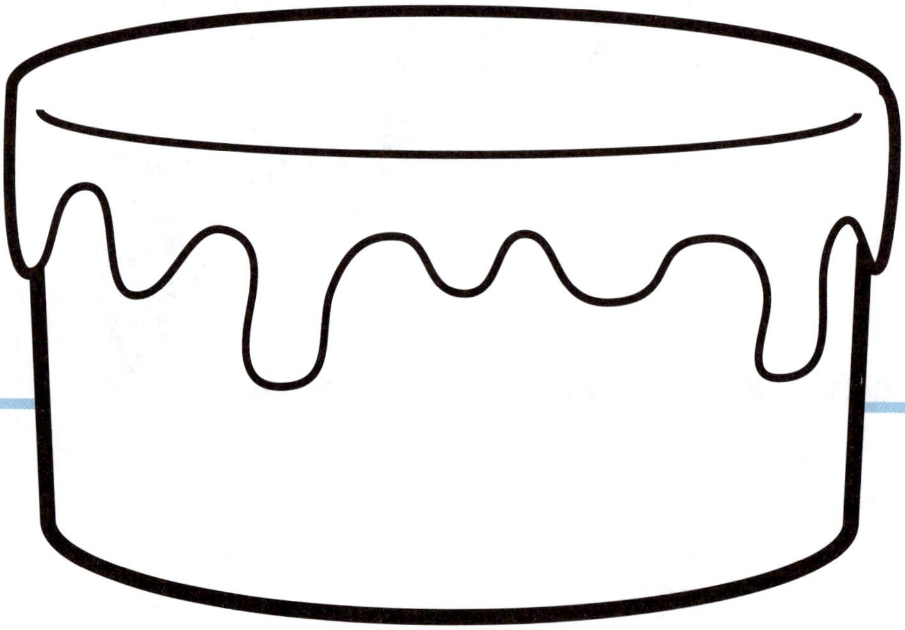

Name _____

12.3

Lungs

What do lungs do?

Air is a gas that has oxygen in it. People use their lungs to breathe air. A <mark>lung</mark> is a part of the body that takes in and lets out air.

God made lungs for people to get the oxygen they need to live. Lungs fill with air when a person breathes in. Air goes out of the lungs when a person breathes out.

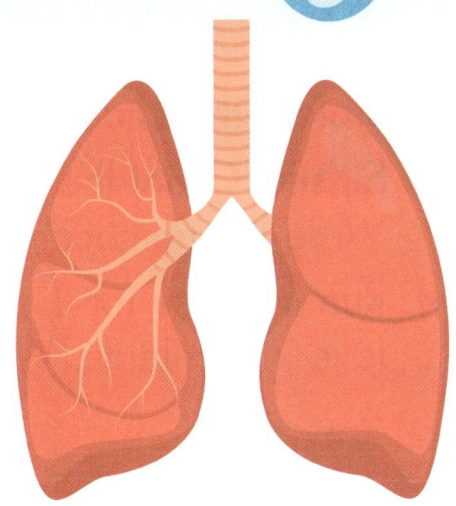

A balloon can be a model for a lung.

1. Circle the balloon that looks like a lung when a person breathes in.

People and other living things need oxygen to live.

2. Underline the groups that need oxygen.

219

Muscles need oxygen to do things. Muscles need a little oxygen when you sit. They need more oxygen when you play.

Do you always take the same number of breaths in one minute?

Complete the chart.

3. Sit still. Count the number of breaths you take in 1 minute. Write the number in the chart.

4. Run in place for 1 minute. Count the number of breaths you take in 1 minute. Write the number in the chart.

	Number of Breaths in One Minute
Sitting	
Running	

Look at the three pictures. Complete the exercises.

5. Make an **X** on the picture that shows muscles working the hardest.

6. The child jumping rope will take **more / fewer** breaths doing this activity.

7. Circle the body part that helps you breathe.

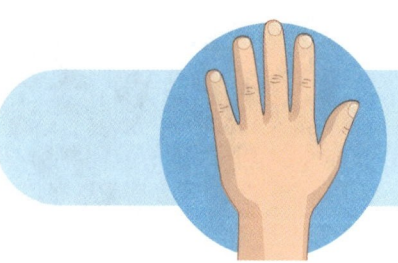
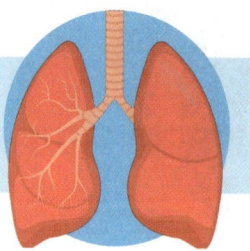
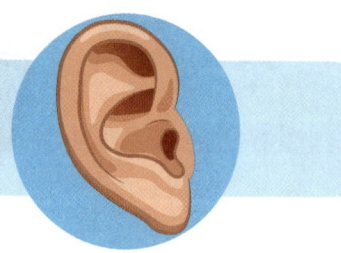

Name _____

12.4

Lungs and Breath

What can you do with your breath?

God made lungs to breathe in and breathe out on their own. God also designed people to be able to use their breath to do different things.

Look at the pictures. Underline the sentences that tell about using breath.

1. The girl sings.

2. The boy ties his shoes.

3. The children fill the balloons.

4. The girl blows bubbles.

5. The child draws on paper.

Circle the correct word to complete the sentence.

6. The girl uses her **blood** / **breath** to sing.

You can make your lungs breathe out to do many things.

The children are doing different things with their breath.

girl and birthday cake

boy in hat and gloves

boy with hot chocolate

girl with flute

7. Choose one picture. Write about how the child is using breath.

_____

_____

8. Draw a picture of something you can do with your breath.

Name _____

12.5

Healthy Heart and Lungs

What should you do to keep your heart and lungs healthy?

A heart needs exercise to stay healthy.

1. Make a check mark for how often you do these exercises.

	every day	sometimes	never
walking			
playing a sport			
moving to music			
riding a bike			
playing outside			

Complete the exercises.

2. My favorite way to exercise is _____

3. Choose one exercise and write a goal to do it often.

_____

A heart also needs healthy food to stay well.

4. Color the path that leads to a healthy heart.

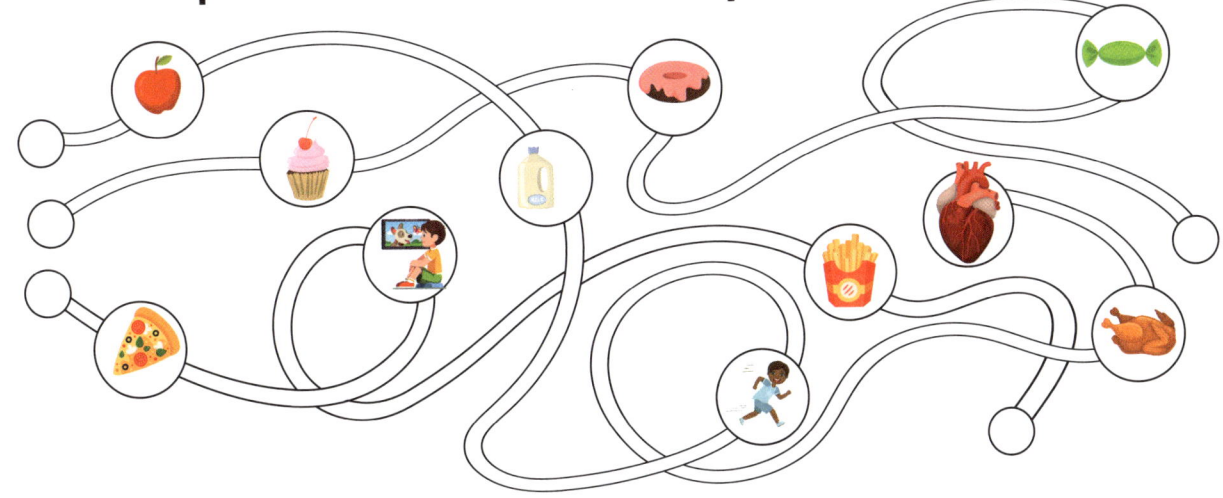

God cares for people. He put tiny hairs in the nose to help keep dirt and germs away from the lungs. Germs are so small that you cannot see them. But they are in the air you breathe. Germs can make you sick.

Help keep your lungs healthy.
When you sneeze or cough, cover your mouth and nose with a tissue. This keeps germs from spreading.

Wash your hands often if you have a cold or if someone around you has a cold.

Circle the correct words to complete the sentences.

5. After you use a tissue, you should **throw it away / save it**.

Then, you should **rub your eyes / wash your hands**.

Name _____

Number of Breaths

Step 1 Ask a Question

_____
_____

Materials
- timer
- _____
- _____
- _____

Step 2 Make a Prediction

I think I will take more breaths

when I _____

than when I _____.

Step 3 Plan and Do a Fair Test

1. Do the first activity for 30 seconds.

2. Count your breaths for 30 seconds.

3. Write the number of breaths. _____

4. Do the second activity for 30 seconds.

5. Count your breaths for 30 seconds.

6. Record the number of breaths. _____

Step 4 Record and Analyze Results

1. my breaths from the first activity ☐ _____

2. my breaths from the second activity ☐ _____   _____

3. Did Activity 1 or 2 use more breaths? _____   − _____

4. Subtract to find how many more.   _____

Step 5 Make a Claim

I took more breaths when I _____ because my muscles needed more _____.

Step 6 Give Evidence

The first activity took _____ breaths and the second activity took _____ breaths.

Step 7 Share Results

1. What did God design the lungs to do when the muscles need more oxygen? _____

2. My number of breaths for the second activity are ___ my partner's number of breaths.
   ○ the same as
   ○ more than
   ○ less than

Name _____

Chapter 12 Review

Use the Answer Bank to complete the sentences.

ANSWER BANK: exercise   scab   lung   carries   heart   heartbeat

1. The _____ is the part of the body that pumps blood.

2. When the heart pumps, or squeezes and relaxes, it makes a _____.

3. Blood _____ things the body needs.

4. A _____ helps a cut heal.

5. _____ helps keep a heart healthy and strong.

6. A _____ is a part of the body that takes in and lets out air.

7. Circle the things that are good for your heart.

Fill in the correct circle to answer the question.

8. What can you use your breath to do?
  ○ warm your hands   ○ wash your hands

9. Why do you breathe in air?
  ○ to stop the spread of germs   ○ to take in oxygen

10. How can you stop the spread of germs?
  ○ cover your mouth when you sneeze   ○ drink a glass of water

Tam finished her math work. She went outside to play for recess. God designed her breathing to change when she plays hard.

Make a check mark next to the answer that tells how her breathing changed.

11. Tam took ___ breaths when she played than when she did math.
  ☐ fewer
  ☐ more
  ☐ the same number of

12. Draw a picture of Tam playing something that makes her take more breaths than she does when sitting at her desk.

13. What is Tam playing? _____

Name _____

Stomach and Food

Chapter 13

Where is this food from?

229

Sushi is from Japan. Foods can come from many places.

Italian food has many pasta dishes.

Mexican food has tacos and tostadas.

American food has hamburgers and hotdogs.

Draw a picture of your favorite food.

Name _____

Food as Fuel

What is food used for?

This truck needs fuel so it can move.

This toy truck needs a battery so it can move.

This fan needs power so it can move.

People need energy to grow and be active. Energy comes from food. All living things need energy. Plants get it from the sun. Animals get it from the plants and animals they eat.

1. Draw a circle around the person getting energy.

2. **Draw yourself eating food that will give you energy for school.**

Good foods give energy. They help the body stay healthy. Milk, yogurt, and cheese make bones stronger. Vegetables, beans, and nuts help the eyes and skin stay healthy. Fruits help the stomach do its job well.

Some foods give energy but are not good for the body. Foods with lots of sugar can make people tired and grumpy.

3. **All these foods give energy. Write them under the body parts they help.**

ANSWER BANK: carrots   pears   cheese sticks   beans   strawberries   yogurt

healthy eyes and skin	strong bones	healthy stomach

4. **Fill in the circle next to the foods that could make you tired and grumpy.**

○ popsicle   ○ green beans   ○ orange   ○ cotton candy

Name _____

13.3

Where Food Comes From

Where does food come from?

People need food to live. Some food comes from plants. God gives plants all they need to grow. He gives plants sun, air, soil, and water to live.

Plants get their energy from the sun. People get some of this energy when they eat food from plants. People eat vegetables, fruits, grains, and nuts that come from plants.

Some food comes from animals. Meat, milk, and eggs come from animals. Some animals eat plants to get energy. People get energy from foods that come from animals.

This cow eats grass to live. People drink milk from the cow.

1. Draw a line from the foods to the plants or animals they came from.

235

Some foods can be mixed to make other food. A recipe gives directions for making food.

Banana muffins are made from these foods. Each food comes from a plant or an animal.

Complete the exercises.

2. Write a food that is used to make banana muffins. Does it come from a plant or an animal?

_____

3. Write a food that comes from a plant and is used in banana muffins.

_____

4. A smoothie may have bananas, strawberries, and milk in it. Where does each of these ingredients come from?

_____

_____

Name _____

Good Food Choices

What makes a healthy plate?

MyPlate is a food guide. It helps you choose healthy food. A healthy plate has food from the vegetable, fruit, grain, protein, and dairy groups.

Where do the foods in these groups come from? Vegetables, fruits, and grains come from plants. Carrots are vegetables. Apples are fruits. Pasta, cereal, and bread come from grains. Beans and nuts are proteins that come from plants. Meat and eggs are proteins. They come from animals. Dairy foods like milk, butter, and cheese also come from animals.

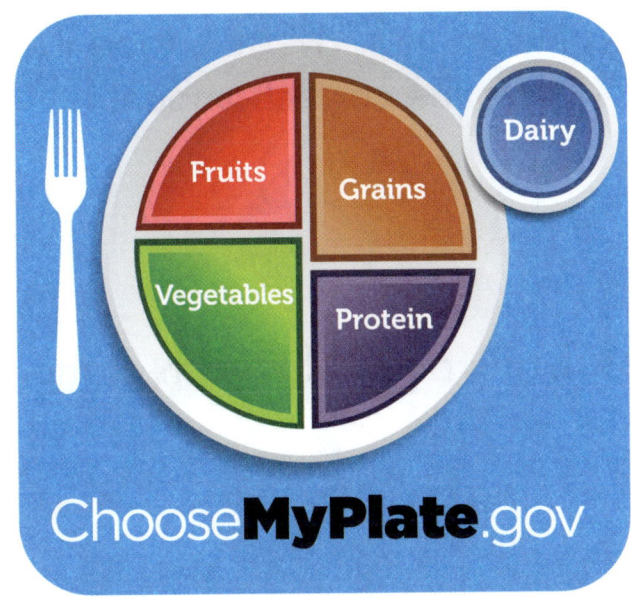

1. Write the MyPlate food group that each picture shows.

The food guide helps you know how much of each food group belongs on your plate. It is better to have food from each group. Too much of just one kind of food is not the best choice.

Complete the exercises.

2. Make an **X** over the food that does not have the correct amount.

3. Draw a food for each group.

dairy	fruits	vegetables	grains	protein

4. Draw a line under the healthier meal.

5. Circle two healthy foods that are part of the less healthy meal.

Name _____

Healthy Habits

What can you do to stay healthy?

A habit is an action done over and over again. Some habits are good. Some habits are not good.

God wants people to have healthy bodies. Good habits make people healthy.

Eating healthy food is a good habit. Drinking plenty of water is also a habit that is good for the body.

Being active by playing or by doing sports keeps people healthy.

1. What can you do today to be active? _____

Healthy habits help people feel better. Here are some more good habits. It is good to do these things each day.

Get lots of sleep at night.

Brush and floss your teeth every day.

Wear sunscreen outside.

2. Make an **X** on the habits that are not healthy. Circle the healthy habits.

Complete the exercises.

3. Write a sentence about one of your healthy habits. _____

_____

4. What habit can you change to be healthier? _____

_____

Name _____

Investigation 13.6

Greasy Potato Chips

Step 1 — Ask a Question

Do chips with more _____ have more grease?

Materials
- 4 baked potato chips
- 4 kettle-fried potato chips
- 2 paper sacks

Step 2 — Make a Prediction

I think Chip ___ will have more grease.

Step 3 — Plan and Do a Fair Test

1. Place the chips from Bag A on Paper Sack A.

2. Place the chips from Bag B on Paper Sack B.

3. Wait 1 hour. Take the chips off both paper sacks.

4. Count the number of grease spots on each paper sack. Write the numbers in the Step 4 table.

Step 4 — Record and Analyze Results

1. Complete the chart.

Compare Fat and Grease

Chip A	Chip B
fat per serving _____	fat per serving _____
number of grease spots _____	number of grease spots _____

2. Draw a picture of the bags and show the grease spots.

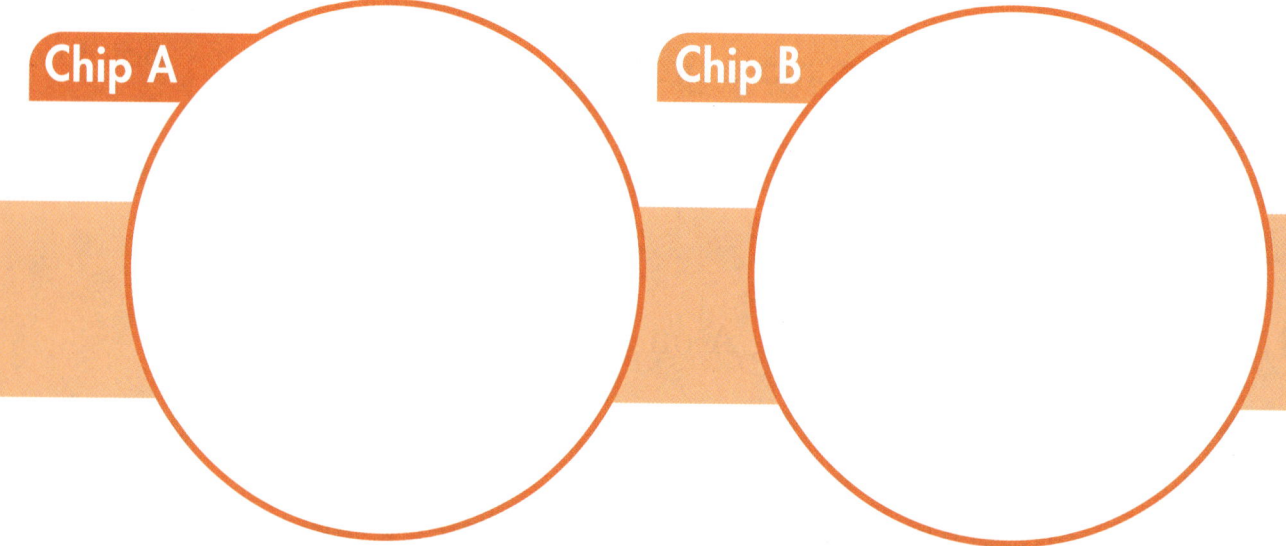

Step 5 — Make a Claim

Chips with more fat have **more** / **fewer** grease spots.

Step 6 — Give Evidence

How could you tell whether chips with more fat have more grease?

_____

Step 7 — Share Results

1. Did other groups have the same results as you did? Yes   No
2. Discuss with your group what you learned about reading food labels. Share how knowing about fat and grease will help you make healthy choices.

Name _____

Chapter 13 Review

Use the Answer Bank to complete the sentences.

ANSWER BANK: tongue   esophagus   stomach   teeth

1. The _____ breaks down food.

2. The _____ tastes food.

3. The _____ moves food from the mouth to the stomach.

4. The _____ chew food.

5. Draw a line from the word to its description.

food •   • something a person does over and over

recipe •   • a set of directions for making food

habit •   • something that gives energy

6. Label the MyPlate food groups.

243

7. Label each food with the MyPlate food group it belongs to.

_____   _____

_____   _____   _____

8. Draw a square around the images that show healthy habits.

Circle Yes or No.

9. God made plants and animals for people to have food.   Yes   No

10. Stomachs have special juices that break down food.   Yes   No

11. Children should get six hours of sleep each night.   Yes   No

12. People should brush their teeth once a day.   Yes   No

13. People need to wear sunscreen in the sun.   Yes   No

14. It is a good habit to know how much food to eat.   Yes   No

Name _____

History

Kinds of Toothbrushes

God gave people teeth for talking, smiling, and eating. It is important for people to clean their teeth to keep them healthy. You may have a toothbrush to help keep your teeth clean. People did not always have toothbrushes.

Look at the different tools people have used to clean teeth.

1. Make an X on the tool that is the most different from the tool you use to clean your teeth.

   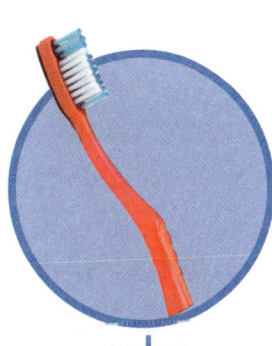 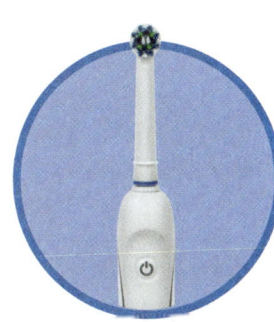

sticks to chew on | hog hair and bamboo | horsehair and bone | nylon bristles | electric toothbrush

Complete the exercises.

2. Do you have an electric or manual toothbrush? _____

3. Draw a picture of your toothbrush.

Unit Connect

245

Technology

X-rays

An X-ray uses a very strong light to make a picture of things that are inside something else. An X-ray only shows one side of an object and does not show color.

Doctors use X-rays to see bones.
Why do doctors need to see bones?

They need to see whether bones are broken.

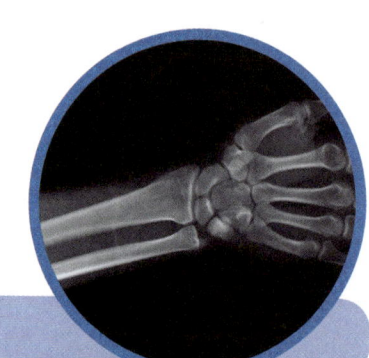

Dentists use X-rays to see inside teeth and gums.
Why do they need to see inside teeth?

They need to see whether teeth have cavities.

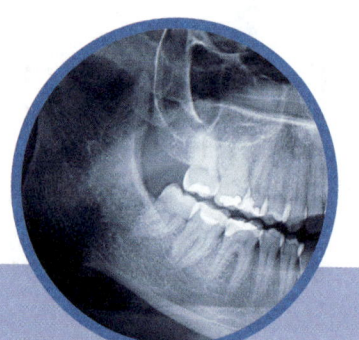

Security officers use X-rays to see inside bags.
Why do they need to see inside bags?

They need to see whether people have items they should not have.

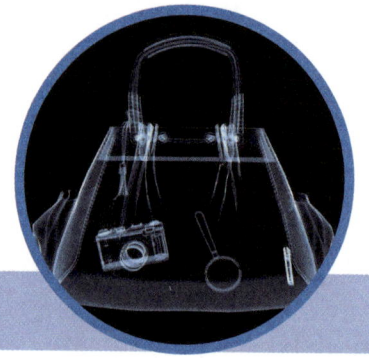

1. What do X-rays not show? _____

2. Have you ever had an X-ray? Why? _____

_____

Name _____

Design

Exercise Course

Exercise helps keep muscles, bones, and the heart healthy. Muscles work harder when you exercise. Working muscles need more oxygen. What fun activities can you do to keep your muscles and heart healthy?

Step 1 State the Problem

Think about the problem you need to solve.
I need two activities that use my legs.
I need two activities that use my arms.

Follow the directions.
1. Circle the activities that only use legs.
2. Underline the activities that only use arms.
3. Draw a rectangle around activities that use both legs and arms.

jumping	hanging from monkey bars	throwing a ball
skipping	walking while bouncing a ball	crab walking
hopping	doing jumping jacks	jumping rope

Step 2 Explore Ideas

Think about what you know about activities.

1. Order the activities to show how hard the muscles work. Order them from least to most work.

   ◯ ___ walking around the block

   ◯ ___ walking to the drinking fountain

   ◯ ___ running a lap on the track

You can move in different directions. You can move side to side, up and down, front to back, round and round, and zigzag.

You can move with one foot or two feet. You can even move using your hands.

2. You usually jump with both feet when you use a jump rope. How could you jump differently?

   _____

3. How can you move with a scooter?

   _____

4. What is one way you can move around cones?

   _____

Name _____

Exercise Course

Step 3 Design a Plan

1. List two activities to exercise your legs.

   _____

   _____

2. List two activities to exercise your arms.

   _____

   _____

Plan to do each activity 10 times.

3. Make a check mark by each movement you will use.

   ☐ side to side   ☐ zigzag
   ☐ front to back   ☐ up and down
   ☐ round and round

4. List the items you will use.

   _____

   _____

5. What is one item you will not use?

   _____

   Why will you not use it?

   _____

6. Use the boxes to draw a picture of each activity in order. Label which muscles are used in each exercise. Draw arrows to show the direction of movement.

7. Which activity might need the most breaths to do 10 times?

Why? _____

_____

_____

Name _____

Exercise Course

Step 4 Build and Test

Use your Step 3 drawings to set up your exercise course.

1. Draw the course after you have set it up. Label the muscles used in each exercise. Draw arrows to show the direction of movement.

2. Will the whole exercise course take a long time or a short time to finish? _____

   Why? _____

   _____

Test your exercise course.

3. Write the number of breaths you take in 30 seconds before doing the exercise course. _____

4. Do each activity 10 times. Do them one right after the other.

5. Write the number of breaths you take in 30 seconds after doing the exercise course. _____

Record your results.

There is enough room to do each activity.	☐ Yes ☐ No
I used my leg muscles with two activities.	☐ Yes ☐ No
I used my arm muscles with two activities.	☐ Yes ☐ No
I took more breaths after the course than I did before the course.	☐ Yes ☐ No

Name _____

Exercise Course

Step 5 Analyze and Redesign

Analyze the results.

1. Did you solve the problem? ☐ Yes ☐ No
2. Did you have fun? ☐ Yes ☐ No

3. What worked best? _____

4. What needs to be changed? _____
   _____

5. How could you change the design so more people could do the exercise course at the same time?

   _____

   _____

Redesign.

6. Draw a picture of your new plan. Label which muscles are used for each activity. Draw arrows to show the direction of movement.

253

Step 6   Share Results

Tell others what you learned from your design.

Underline all the words that make a correct statement.

1. When I exercise my legs and arms, I breathe **more / less**.

   Exercise helps keep my **heart / teeth / muscles** healthy.

Complete the exercises.

2. Which exercise did you like best for your legs? Why?

   _____

   _____

3. Which exercise did you like best for your arms? Why?

   _____

   _____

Circle the correct word to complete the sentence.

4. I know that God is wise because He makes my **lungs / mouth** breathe.

5. I breathe in the **oxygen / blood** I need for the activity I am doing.

254

Glossary

B

beak — the hard part of a bird's mouth

body — the whole physical self

bone — a hard part of the body that gives the body shape and helps it move

D

dentist — a doctor who cares for teeth

E

environment — all the living and nonliving things in an area

F

fin — a part of a fish that helps it swim

H

habit — an action a person does over and over again

heart — the part of the body that pumps blood

heartbeat — the squeezing and relaxing of the heart muscle

J

joint — a part of the body where bones meet

L

leaf — the part of a plant that makes food

life cycle — the series of stages in the lives of plants, animals, and people

light — a type of energy that makes seeing things possible

living — grows and changes and needs air, food, and water

lung — a part of the body that takes in and lets out air

M

mammal — an animal that has hair and can make milk to feed its babies

measure — to find the size, length, or amount of something

N

nonliving — does not grow or need air, food, or water

P

parent — an animal that has babies

predator — an animal that hunts another animal for food

pull — to bring something toward you

push — to move something away from you

R

recipe — a set of directions for making food

root   the part of a plant that holds the plant in the ground

S

safety   freedom from being hurt

season   a time of year

seed   the first stage in a plant's life cycle

senses   ways the body observes and learns about the world

shadow   an area of darkness created when a source of light is blocked

star   an object in the sky that makes its own light

stem   the part of a plant that holds up the plant

stomach   a part of the body that breaks down food

sunlight   light that comes from the sun

T

telescope   a tool that makes things that are far away appear closer and larger

thermometer   a tool used to measure how hot or cold something is

tool   any object used to do work

V

vibrate   to move back and forth very quickly

W

weather   what the air is like outside

Y

young   an animal in an early stage of life

Science Journal

Science Journal

Science Journal

Science Journal

Science Journal

Science Journal

Science Journal

Science Journal

Science Journal

Science Journal